COLORADO'S
Mrs. Captain Ellen Jack

MINING QUEEN OF THE ROCKIES

JANE M. BARDAL

THE
History
PRESS

Published by The History Press
Charleston, SC
www.historypress.com

First published 2023

Manufactured in the United States

ISBN 9781467153638

Library of Congress Control Number: 2022950050

To Mom and Dad.

CONTENTS

ACKNOWLEDGEMENTS

I have many people to thank who helped me with some part of this book. Virgil Lueth, an organizer of the New Mexico Mineral Symposium, accepted my two proposals to talk about Captain Jack. The first talk I gave there in 2013 started me on the journey to discovering her story. Mark Jacobson helped me find information about how to access records of mining claims.

The Charles Redd Center at Brigham Young University provided me with a travel grant in 2016 that paid for my travel expenses to conduct research in Aspen, Gunnison and Denver that year.

The staff at many places helped me locate valuable original documents: the Colorado State Archives; the Bureau of Land Management office in Golden; the National Archives at Denver; and the clerk and recorder's offices in El Paso, Teller, Ouray, Saguache and Pitkin Counties. I would like to give a special thank-you to the Gunnison County clerk of court Betsy Nesbitt and Kathy Simillion, Diane Folowell and other staff members at the Gunnison County Clerk and Recorder's Office, who gave me access to numerous court records. Thank you to Jane Holmes, the clerk of court of Ouray County, who also helped track down court records. Many librarians also helped me: those at Western State College in Gunnison, the Hart Research Center at History Colorado, the School of Mines at Golden and the Denver Public Library–Western History Resources and Lisa Kindrick at the Albuquerque Public Library.

I presented talks about Captain Jack at several meetings: the Western Social Science Conference (2014), the Mining History Conference (2018) and the Pikes Peak Regional History Symposium (2019). Thanks to the organizers and audiences, who provided further encouragement with the project.

Dianne Layden and Hana Norton provided helpful comments on the manuscript. Thanks to many other friends who listened to me as I discovered the many episodes about Captain Jack.

INTRODUCTION

*E*llen Jack made a surprise trip to her Black Queen Mine, located on the steep slope of Sheep Mountain, in a remote area southwest of Aspen, Colorado. She found that the miners were trying to cheat her. She told one of the miners, Mr. Aller, "You cannot move that mineral."

He said in a sneering way, "And who will stop me?"

Ellen said, "I will."

He laughed in Ellen's face. She went down the trail to the cabin on the adjoining Fargo claim and said to the three men, "Lend me your rifles and your shotguns."

She returned to the mine, well-armed and ready to fight to keep her sacks of silver ore. Mr. Aller approached the mine, and when he saw Ellen, he turned white as death and said, "What in hell are you doing here?"

She replied. "I am on my own ground and you are a thief, and I have a right to protect my property. Your bond and lease are forfeited. You get off this property."

Ellen stopped the burros that were coming up the trail. Ellen issued an order to the burros' owner: "Mr. Benton, turn your jacks off this property, for not one sack of mineral leaves this place."

Mr. Benton tried to pull his gun out of his belt. Ellen said that she "sent a shot and took his ear off as clean as though it had been cut off and was going to send another when he threw up his hands and yelled out, 'I am shot.'"

Ellen Jack may have added this detail of shooting off an ear for dramatic effect, as that part of the story did not appear in any news reports. But the

Mrs. Captain Jack, mining queen and dead shot. *Author's collection.*

gist of the story appeared in several newspapers around the country, with the headline in Denver's *Rocky Mountain News* stating: "Mrs. Jack's Shot Gun: Mrs. Ellen Jack Compels a Jack Train Loaded with Ore to Unload at the Muzzle of a Shot Gun." Ellen Jack won this round of fighting, but in trying to profit from the mine, she would have to engage in many more battles against lawyers, the sheriff and capitalists who would try to squeeze

Mrs. Captain Jack, mining queen of the Rockies. *Author's collection.*

it from her hands. She engaged in shenanigans herself in attempting to gain wealth from the mine.

After prospecting and owning mines on the Western Slope of Colorado in the 1880s and 1890s, Ellen Jack staked mining claims in the foothills west of Colorado Springs in the early 1900s. She ran a curio shop and restaurant that catered to tourists. She sold postcards in which she proclaimed the self-described title "Mrs. Captain Jack, mining queen of the Rockies." Postcards in the early 1900s often showed towns, scenery and mining operations, but it was unusual for a woman to publish a set of postcards of herself. In promoting her story and image, Captain Jack drew on and contributed to the popular mythologizing of finding one's fortune in Colorado during its pioneer days.

Ellen Jack established her legacy as a Colorado pioneer by publishing an autobiography in 1910, *The Fate of a Fairy, or Twenty-Seven Years in the Far West.* One tourist who visited Captain Jack's cabin on the High Drive said that Ellen referred to herself as "the Fairy," which can refer to "a woman thought to possess extraordinary or magical powers, an attractive or seductive woman, or a small or delicate person." Ellen described herself as a spiritual medium and an attractive woman. She was petite in stature, probably not more than five feet tall, although she was certainly not delicate in terms of her personality.

Most of the stories she told are true, but she did tell some tall tales and left out a lot of other information. This biography will use other historical sources, such as newspaper articles and court records, to tell her life story more fully. I take her autobiography as a record of her perspective, and I will note where there are significant differences between her account of events and other sources.

This biography focuses on Ellen Jack's search for gold and silver in Colorado and her self-promotion as the "mining queen of the Rockies." In addition to being an account of her many mining adventures, this biography is an even richer story of a persistent woman who followed her own path, a creative woman who wriggled her way out of predicaments in unexpected ways, a feisty woman who struggled with those around her in ways that were not always pretty and an independent woman who followed her own inner light throughout her life.

1

ELLEN IS DESTINED TO FIND HIDDEN TREASURES

Born in the original Fox homestead in New Lentern, Nottingham, England, Ellen Elliott would follow in the Quaker tradition by expressing idiosyncratic views on religion and challenging authority. George Fox founded the Quakers in the mid-1600s, and he criticized religious and political leaders. Quakers believed that the spirit of God lay within each individual and that each person should follow the authority of their own inner light.

Ellen's quest for minerals was prophesied by a fortune teller when she was a young girl. The fortune teller told Ellen's mother that "this child was born to be a great traveler, and if she had been a male would have been a great mining expert. She is a Rosicrusian [*sic*], born to find hidden treasures. She will meet great sorrows and be a widow early in life."

The fortune teller's prophecies came to pass. Ellen Elliott traveled to the United States with her sister, and on the return voyage, she met Charles Edward Jack, the first officer of the ship. After their marriage in 1860 in England, the couple resided in Brooklyn, New York. Ellen Jack became a widow after Charles died in 1874 at the age of forty-three from injuries he sustained during his service as a naval officer in the Civil War. Charles had been called "Captain Jack," and after her husband's death, Ellen referred to herself with that title.

Ellen Jack obtained a widow's pension, which provided her with a degree of financial independence to pursue her own goals. Congress had enacted the Widows Pension Act of 1862 to encourage married men to enlist in the military so that their families would not become destitute in the event of their deaths. Ellen's pension of $264 per year was substantial but not an extravagant amount of money when compared to an average worker's wages of $438 per year.

Further misfortune propelled Ellen to embark on a different course of action in her life. Two of her children had already succumbed to scarlet fever, and in 1878, her daughter Margaret also died from the dreaded disease. Ellen became overwhelmed with grief. She became acquainted with four women spiritualists. Believers in spiritualism claimed to be able to communicate with the dead. Ellen agreed with this idea, stating, "I was satisfied in my own mind that the spirit could come back to this world again at some time or other and that we have spirits around us all the time—either good or evil—and that they influence us." The women spiritualists identified Ellen as a healer, and one of the women, Madame Clifford, asked Ellen to care for her child, who was dying of Bright's disease. Madame Clifford, who advertised herself as the "great American medical and business clairvoyant and prophetess," echoed the fortune teller's prophesy that Ellen was a Rosicrucian and that she was born to find hidden treasures.

Ellen asked her sister-in-law, Margaret Quackenbos, to look after her ten-year-old daughter, Adeline. Ellen described Margaret as "beautiful…with a heart as cold as ice." Ellen's decision to leave Adeline behind was quite unusual for a mother, and Ellen would have a contentious relationship with her daughter Adeline for the rest of her life.

At the age of thirty-seven, Ellen Jack set out for Denver to find her fortune. Prospectors had rushed to the area over two decades earlier. The argonauts had soon exhausted the placer gold in the streambeds, and then new mineral discoveries to the west of Denver ensured the city's place as a gateway to the riches of the Rockies. Ellen probably arrived via one of the two railroads that had reached the city by 1870. Colorado was granted statehood in 1876, with Denver as the capital, and by the time of Ellen's arrival in 1880, the city had a population of 35,629.

Shortly after Ellen's arrival in Denver, she ran into someone who would help point her in the direction of the hidden treasure she sought. While walking down a street that was lined with newly erected brick buildings, she heard a woman call out, "Captain Jack!" She turned to find a well-dressed woman, who, years before, had been her old nursemaid. Jennie told Ellen that she had married but that her husband beat her and frequented prostitutes, so she left him. Jennie entered Madame Clara Dumont's house on Holladay Street in the red-light district, and she vowed to "wreck the life of every man she could to avenge the wrong that had been done to her." She opened her own sporting house, and men came to her with their money: "When their cash is gone, they are politely shown to the door."

Could this Jennie have been the famous madam Jennie Rogers? Jennie Rogers arrived in Denver and bought a house on Holladay Street in January 1880, so she would have been in Denver at the time of Ellen's arrival. Jennie Rogers ran her business in the vicinity of Holladay Street for over two decades.

Jennie encouraged Ellen to go to Gunnison. Early reports about the Gunnison Country extolled it as the "land of promise," and it was "all the rage" with its "unclaimed bonanzas." Denver's *Rocky Mountain News* exclaimed that "no doubt mines will be worked in this district for the next hundred years successfully, as they are in unlimited quantity, and, out of so many, there must be some of great richness." Ellen set out for Gunnison to find hidden treasure, and she would find plenty of other business opportunities there as well.

Ellen said that one of her fellow passengers on the stagecoach was a future governor of Colorado, Alva Adams, who planned to open a hardware store in Gunnison. At one of the stops, the passengers found out that a previous stage had been robbed. Alva Adams packed guns in his two large satchels. He devised a plan with the other men in case they encountered the robbers. To Ellen, he said, "Do not get frightened at the sight of those guns, but we must be prepared." She replied, "I came prepared," as she took out her Smith & Wesson .44-caliber revolver.

Years later, Ellen would tell Anne Ellis that the stage had been held up. In 1929, Ellis wrote her autobiography about living in several mining camps in Colorado during the same era as Captain Jack. Ellis wrote, "Cap. Jack shot right along with the men, only better, as she could and did clip a finger or an ear off at will."

When a woman at a hotel in the town of Saguache questioned Ellen about traveling alone as a woman, Ellen replied, "I do not fear man or devil; it is not in my blood, and if they can shoot any straighter or quicker than I, let them try it, for a .44 equalizes frail woman and brute man, and all women ought to be able to protect themselves against such ruffians."

From Saguache, the stage continued up broad valleys and over the forested Cochetopa Pass at an elevation near ten thousand feet. Ellen would stake gold mining claims in this area in the 1890s. The next day, the stage stopped in Parlin, where Ellen met the handsome man who would become her husband and business partner in a few months: Jeff Mickey. The travelers rode the final dozen miles along Tomichi Creek into Gunnison.

The town took its name from Captain John Williams Gunnison, who had explored the area for a transcontinental railroad route. Prospectors had made some discoveries in the Gunnison Country in 1861, but Native attacks sent them back over the range and kept them out for over a decade. The U.S. government made a treaty with the Ute people, which pushed them farther west. Sylvester Richardson formed a colony to settle this area in 1874. He foresaw that the sagebrush plain near the junction of Tomichi Creek and the Gunnison River was situated like the hub of a wheel, with streams flowing out of the mineral-rich mountains to this central point. At the time of Ellen's arrival in May 1880, Gunnison comprised around fifty buildings and nearly as many tents.

Ellen opened a restaurant and boardinghouse. She had experience in the hospitality business, as she ran the Bon Ton Hotel on Coney Island. Despite the hotel's elegant name, the *Brooklyn Daily Eagle* described the Bon Ton as a "somewhat unsightly hotel," which "has not borne an enviable reputation" since its opening. Fire destroyed the Bon Ton Hotel in March 1876, and despite accusations of arson, Captain Jack got some insurance money for her losses after three years of court cases. Ellen lost her business, and she was

MAIN ST. LOOKING NORTH, GUNNISON, COLORADO.

Jack's cabin was one block east from this intersection of Main Street and Tomichi Avenue. *Author's collection.*

Gunnison Courthouse, circa 1910. *Author's collection.*

bitter that her friends now avoided her. She said of this experience, "I began to see that the only friend on Earth was money, and not only a friend, but power; that I must stir and do something or go somewhere."

Now, Ellen would run another establishment that would bear a similar reputation. Located along the major artery of Tomichi Avenue, one block east of Main Street, her boardinghouse was called "Jack's Cabin." Opposite from Jack's Cabin, George Walsh ran the Q.T. Saloon, one of many such establishments in the new town. One block to the north, a two-story brick courthouse would be built over the summer. The location of the courthouse would be quite convenient for Ellen, as she would become a frequent visitor in the next few years.

Ellen Jack's boardinghouse should not be confused with another "Jack's Cabin" that is located several miles north of Gunnison along the road to Crested Butte. At this location, Jack Howe ran a grocery store, post office, hotel and saloon that served freighters and other travelers.

Ellen bought materials for her boardinghouse from a man who had lost money in a keno game. The twenty-four-foot-by-fifty-foot tent had "six small rooms partitioned off for bedrooms for the girls." Ellen did not explicitly mention whether the girls were prostitutes or not, but in publishing her autobiography, it would have been wise to omit any mention of illegal activity. On the other hand, author Betty Wallace, who published *Gunnison Country* in 1960, said that Captain Jack was "no floozy." Young boys would send "men seeking a fast house over to Mrs. Jack, just to hear the explosion when they approached her with their ideas."

Ellen was in a good position to hear the latest news and information about mining discoveries at Jack's Cabin, because many freighters passed through town on their way to the surrounding camps. Several other women who ran boardinghouses in the West became involved in mining after gaining information from the men who boarded with them. With all the news and rumors of mining strikes swirling around Gunnison in the summer of 1880, it looked like Ellen would be able to share in the excitement when she learned of a secret gold deposit located southwest of Gunnison. She heard about it from Marshal John Roberts, who had gotten the information from Bill Edwards.

Edwards had a reputation for being a tough character. In March 1880, he was arrested for assaulting a respectable citizen, and he had been fined for drunk and disorderly conduct and carrying a concealed weapon. The *Gunnison News* opined that "such a willfully vicious character should be made an example of." In mid-August, Marshal Roberts arrested Edwards for drunk and disorderly conduct. Edwards owed a fine of forty-two dollars for the charges against him. John Roberts asked Ellen Jack if she would pay the fine in exchange for a promise to cut her in on a share of the gold discovery. She agreed to the deal.

Edwards revealed that he had lived with the Utes for years and that they had told him of a gold deposit. They gave him permission to mine it, with the stipulation that he not tell anyone else about it. The Utes threatened to kill any trespassers. Edwards went against this prohibition by showing Roberts the location. Shortly after his release from jail, Edwards was hanged for horse stealing. Ellen and John Roberts could not get to the mine because it was on the Ute reservation. They would have to wait until they could gain access to that area before they could go in search of great riches.

In October, Jeff Mickey opened a saloon on Ellen's property that was also called "Jack's Cabin." The establishment soon gained a bad reputation as a place characterized by drunkenness and fighting. Thomas "Jeff" Mickey, age thirty-two, had been a dry goods merchant in Shelby, Ohio, and he had one of the finest private libraries in the state. His father, also named Thomas Mickey, had been a wealthy businessman and a member of the Presbyterian Church. Jeff had married Ida Wilson in 1867, and they had a daughter the following year. Jeff went to Coffeyville, Kansas, and in April 1879, he headed farther west from Kansas to Colorado. His wife, Ida, died in June of that year in Ohio, and his daughter went to live with her grandparents. Regarded by some as a quarrelsome man, Jeff had survived a gunshot wound that was inflicted during a fight with another man. In Gunnison, Jeff was regarded as a good man when sober, but he got into trouble when he drank.

On Christmas Day, the Methodists first rang their new church bell, which had been donated to them by Elisha Buck, the owner of the *News-Democrat*. The newspaper expressed the hope that the bell's tolling would serve as a warning, "conveying to many thoughtless ones that it is the Lord's Day." But at Jack's Cabin, the revelers failed to heed the warning, and an affray broke out that evening.

Several Frenchmen had arrived in town to celebrate; they drank too much liquor and refused to pay their bill. Harsh words escalated into a fight that spilled out onto the street, breaking the stillness of the silent night. Captain Jack joined the row and attempted to assault one man with a bottle, but she was no match for the men, who beat her with clubs, sticks and their fists. Three shots rang out from saloonkeeper Jeff Mickey's gun, and one bullet burrowed into Joseph Clemens's right arm, below the elbow. Jeff Mickey suffered a severe beating, and the assailants landed in jail.

Ellen sustained an injury to her forehead that left a permanent scar. Ellen Jack's assailant, Joseph Clemens, faced a judge in mid-January 1881, and he received a fifty-dollar fine for assault. The beating left Ellen bruised and sore. She needed a doctor's care for her wounds, and she suffered a partial loss of vision. Incapacitated for the next six weeks, she was unable to carry out her normal business of running the boardinghouse. Thomas "Jeff" Mickey faced charges of assault with intent to commit murder. Ellen and three men posted a bond for him, and he was released on his own recognizance. Joseph Clemens recovered from his bullet wound.

Ellen Jack and Jeff Mickey were married in March, according to a brief notation in court documents in this assault case, but there are no other official records of the marriage. In her autobiography, Ellen told a different story: Jeff asked her to marry him, but she refused because she did not like his habit of running around with sporting women.

Once she was back on her feet again, Ellen filed a civil suit for monetary damages against her assailants, Joseph Clemens, Frank Clemens and Frank Seymour. She charged that these men, "with sticks, clubs, and other deadly and dangerous weapons, and with their fists, without any cause, provocation, or excuse, did violently assault, beat, and seriously wound, injure, disable and disfigure her with blows to her forehead, face, and body." She sought damages of $2,000 for doctors' bills and being unable to transact business during her recovery.

The assault case was heard before a six-man jury in Judge David Smith's courtroom in April 1881. Ellen called two doctors and five other witnesses to testify on her behalf. Ellen's lawyer argued that Joseph Clemens had admitted to assaulting Ellen, so the jury should issue a finding in her favor. The jury had the discretion to award additional damages above the $2,000 requested to punish the defendants for their "violation of personal rights and social order."

Judge Smith's instructions to the jury on behalf of the defendants came close to telling the jury what their decision should be. The judge argued that even though Joseph Clemens had struck Ellen Jack, the jury should find in favor of the defendants, as Ellen had come out of her house "to the scene of the affray and had taken a part in it." She had assaulted the defendants with a bottle and had sustained her injuries while taking an active part in the fighting. The jury had to decide if she "was entirely blameless in the matter" and if "the injury was without any fault" on her part. Judge Smith also stated that the fact that Ellen Jack was a woman should not carry any weight in their deliberations.

The judgment went against Ellen Jack. She paid fifty dollars in court costs. Meanwhile, in the May 1881 term of the district court, Thomas Jeff Mickey was found guilty of assault with the intent to commit murder against Joseph Clemens. He paid an eighty-dollar fine.

Also in mid-May, Sheriff George Yule arrested both Thomas Mickey and Ellen Jack on charges of keeping a disorderly house. The complaint alleged that on December 25 and on many other days in the ensuing months, the couple permitted and encouraged gaming, drinking, cursing, swearing, quarreling, challenging to fight, fighting and otherwise misbehaving during both day and nighttime hours. Their conviction resulted in a thirty-dollar fine.

In her autobiography, Ellen recounted these events quite differently from newspaper accounts and court records. Ellen told a story of how Frenchmen and a band of Natives came into Gunnison to "take the town." During the attack, a Native struck her in the head with a blow from a poisoned tomahawk. While Ellen lay wounded and suffering from the toxic blow, the Ute chief Colorow heard of her plight, and he fondly remembered the woman with the golden locks of hair. He arrived at her bedside with a medicine to counteract the poison.

It is unlikely that this melodramatic story is true. No reports describe the Utes attacking Gunnison, although residents feared an attack during the early years of the town. In addition, Colorow hated white settlers because of their takeover of Ute lands, making him an unlikely hero to save a white woman. Many white Coloradoans expressed a virulent hatred of Colorow and the Utes.

IN THE SUMMER OF 1881, Jack's Cabin and many other Gunnison businesses boomed, and merchants made fortunes by providing lodging, meals and entertainment. The influx of people swelled the town's population to 4,000. Brick, stone and wood frame buildings replaced the tents that housed many early businesses. A two-story brick schoolhouse served 180 pupils, and seven churches ministered to the faithful. The Denver and Rio Grande Railroad made its first appearance in August and promised to bring additional development and prosperity to the Gunnison Country.

N.P. Babcock wrote frequently about Ellen Jack and her companions in the *Gunnison News-Democrat*. The newspaper's owners had a financial interest in promoting the Gunnison Country because they had invested in numerous town lots and nearby mines. Babcock had been trained at a fast-paced New York City newspaper, so he thought that the position of editing a small-town newspaper would afford him plenty of time for relaxation and reflection. But instead, he found himself to be as busy as a reporter covering the Bowery, an area of New York City known for its brothels, bars and flophouses. Years later, Babcock would refer to himself as a tenderfoot who became disillusioned by the frequency of gunplay, gambling by both men and women in the front rooms of all-night saloons and a lynching.

A competing newspaper, the *Gunnison Review*, took plenty of jabs at Babcock, describing him as the "tall darning needle," in reference to his tall and thin build, as well as calling him a series of inventive names: "a petticoated tenderfoot," "that old lady in charge of Buck's pocket edition" and "that lavender-haired contemporary who is so unprofessional, uncouth, and devoid of merit." While the *Review* reviled Babcock, his dramatic reporting style preserved a colorful rendition of Gunnison's early days.

Babcock's handiwork appeared in a headline in the *Gunnison Daily Democrat* that declared: "Mark Twain Outdone: The Most Gigantic Joke of the Season at the Expense of Mortality." At Jeff Mickey's saloon, a railroad worker who had been drinking for several days suddenly dropped dead of an apparent heart attack. Some people speculated that the large amount of whiskey might have had something to do with his death. His body was laid out on top of a table in the saloon, with a sheet draped over it, and candles were stuck in the tops of beer bottles and placed around his head and feet. This odd sight attracted even more customers than providing a free lunch or a brass band.

In addition to a restaurant and rooms for boarding, Jack's Cabin offered a variety of entertainment acts. A comedian named Murphy performed sketches, and Jeff Mickey provided a turkey lunch for the July 4 festivities. Jeff Mickey expanded his business by building a gymnasium, boxing school and bowling alley next to his saloon. A boxing match and music entertained the large crowd on the opening night in September.

Gunnison's population swelled in the late fall of 1881, as many people left their high mountain mining camps before they became snowbound. They would spend the winter and early spring in town, and with little to do, trouble was sure to follow.

Jeff had known Jimmy McLees in Coffeyville, Kansas, and in 1879, they headed west with a party of several other men. Jimmy McLees, sometimes referred to as "Old Man McLees," due to his age of about fifty years, stood five feet, eight inches tall and walked slowly with his toes turned out. He had a light moustache and thinning hair, a red face and watery, gray-colored eyes. By all reports, Officer McLees performed his duties well.

On the night of Monday, December 19, Officer Jimmy McLees shot saloon owner Cal Hayzes. McLees entered the saloon at midnight to arrest one of Hayzes's employees. Hayzes ordered McLees to get out, and he grabbed McLees by the arm and shoved him toward a side door. After a few steps, Hayzes let go of his arm, and the two men faced each other, standing a few feet apart. Both men pulled out their revolvers and fired. Hayzes's shot missed, but McLees's shot hit its mark. Hayzes dropped to the floor. Hayzes died on Wednesday morning due to the infection that developed from the bullet wound. His dying wish was for McLees to be brought to justice.

Jimmy McLees was arrested on a charge of murder. McLees was not locked up in a jail cell but was placed under guard at the sheriff's office. The guard locked the door to the office, and when the guard and McLees retired for the night, the guard placed the keys in his pants pocket and tucked his pants under his pillow. The guard awoke around 4:00 a.m. on Friday morning to find McLees and his pistol gone.

The guard notified Undersheriff Clark, and they immediately began searching for McLees. Rumors flew concerning the whereabouts of McLees. Some people speculated that he had been provided with a horse and a

Winchester and that he had raced out of town toward the Black Canyon. Others thought that he might still be in town. A friend warned that McLees had said that he would rather die than be taken back to jail. Sheriff George Yule offered a $200 reward for his capture.

On the afternoon of Sunday, January 1, Sheriff Yule, Undersheriff Clark and a posse of eight other men made an unannounced appearance at Jack's Cabin. Jeff Mickey stood behind the bar, serving a few customers. The sheriff secured Jeff's guns. Guards stood at the doorways to prevent anyone from entering or leaving. The sheriff and his men went to Mr. and Mrs. Mickey's bedroom, where Clark moved two trunks.

Ellen held out a set of keys and said to him, "Look in all the closets of the house if you please, but don't pull my bedroom to pieces."

Clark replied, "We propose to look all through the house, but in the first place, we will make a thorough search of this room."

Two of the men pulled up a carpet to reveal a trapdoor. They opened the trapdoor but couldn't see anything below due to the darkness. Ellen called Jim's name several times but got no reply. Finally, she said, "There is no use Jim; there are fifty men here with guns, and you might as well come out without losing your life or shedding their blood."

Jim relented, "I will come out, but I am afraid of Sheriff Clark."

Clark assured him, "If you come out quietly and without any further trouble, you will not be hurt."

McLees crawled out of the cellar, and he was taken to the courthouse, where this time, he was secured in the locked jail cell. McLees explained to a newspaper reporter who interviewed him in his jail cell that he had left the jail because he was sick. He went to Jack's Cabin because that was where he lodged. He asked Mrs. Ellen Mickey to hide him until he could get his bond money. He had no intention of fleeing. He felt confident that he would be acquitted at a trial because he had performed his duties as an officer and had shot Hayzes in self-defense.

Sheriff Clark, Judge Florida and two other men returned to Jack's Cabin to arrest Jeff and Ellen Mickey for harboring a fugitive. Upon entering the dining room, Clark pointed a shotgun at Jeff and ordered him to put his hands up. A search of his person revealed a pistol. Clark took Jeff to jail and put him in the cage with McLees and nine other inmates. Ellen was told to appear in court the next morning. The primitive jail facilities consisted of a single cage, with no separate facilities for a woman.

Around midnight, Ellen swore out a warrant before Judge Florida for the arrest of Sheriff Clark on a charge of false imprisonment. As Judge Florida walked toward the courthouse to serve the warrant, he became suspicious at the sight of thirty or forty men assembled at the courthouse. He warned them to get back.

Fears of lynching were justified. At the end of October, an Italian worker who had been charged with killing his boss was forcibly taken out of jail by a group of masked men and dragged through the snowy street with a rope tied around his neck. By the time they strung him up on a livery sign on Tomichi Avenue, he was already dead.

A constable knocked on Sheriff Clark's door, awakened him and served the warrant. Clark refused to give up the keys to the jail, and the judge agreed that the issue could wait until the next morning. Meanwhile, a heavily armed Sheriff Yule ran to the courthouse and dispersed the crowd without further incident. Ellen failed in her attempt to have Sheriff Clark charged with false imprisonment.

Jimmy McLees stayed in jail until Tuesday, January 24, when he finally got enough money for his bond. To celebrate his release from jail, Jimmy went on a drinking spree with Jeff. Jeff and Ellen had a good marriage under normal circumstances, but she became angry when Jeff drank excessively. The next day, the couple exchanged hot words. On Thursday, Ellen left for Crested Butte to consider leasing the Miners' Boarding House. She hoped that getting Jeff away from Gunnison would be good for them. She also hoped that her departure would bring him to his senses and that he would quit drinking.

Around midnight, Jimmy went into the barroom, where Jeff was sitting. Jeff held up a bottle of morphine and said, "Here's the thing that will end all my troubles."

Jimmy asked, "Do you intend to take it?"

"Yes."

Jimmy thought he was bluffing. "You ain't going to take that damn stuff. Go and lie down, and when you get sober, your wife will come back. I'll run the place while you're asleep."

"All right. I'll do that," Jeff replied. Jimmy reached for the bottle, but Jeff jerked it away and refused to give it to him. Jeff went to his bedroom, but just a few minutes later, he called out to Jimmy, "Mac! Mac! Mac!" Jimmy rushed to his bedroom.

"Well, I've done it," Jeff said as he stood in the doorway crying. Jimmy saw the empty bottle on the table. Jeff had taken fifteen grains of morphine, enough to kill any man, especially when combined with alcohol. A one-quarter grain of morphine was the typical dose for medicinal purposes.

Jimmy sent for Dr. Rockefeller immediately. Jeff lapsed into unconsciousness thirty minutes later.

A telegram sent to Ellen in Crested Butte contained a critical mistake. It was supposed to say, "Jeff has taken poison," but instead, it read, "Jeff has taken horses." Ellen thought that Jeff would be coming up there the next day with the horses. She did not realize the severity of the situation.

Dr. Rockefeller made many attempts to revive Jeff as he lay unconscious on his bed in the cramped bedroom that was barely lit with a small lamp. He tried emetics, to no avail. At one point, Jeff appeared to stop breathing, so for two hours, Dr. Rockefeller applied pressure to his chest at regular intervals to restore his breathing. The doctor beat Jeff's bare feet with a pine board. He used a battery to apply electricity to his chest. A slap on the cheek resulted in Jeff opening his eyes for a few moments. Jimmy pleaded, "Wake up, Jeff, wake up."

N.P. Babcock, the editor of the *Gunnison News-Democrat*, described McLees's reaction to Jeff's impending death:

> *A searcher after bits of western romance need have gone no further for his supply. As tenderly as the young mother watches the illness of her first born, this man McLees, who has been described as a type of all that is hard and brutal; this man, who is known to carry in his flesh the lead of half a dozen bullets; this man, who but a month ago riddled a man with pistol balls, and showed no sorrow for the act; gazed upon the face of his unconscious friend, and hardly knowing what he did, fondled his feet, and called upon the doctors to save him.*

At 3:00 a.m., a reporter asked Jimmy if he thought Jeff would recover. He answered, "Of course he will. Do you suppose Jeff Mickey was ever made to go that way? Look at him—me and him, one day on the banks of the North Canadian, stood off a dozen Kickapoo Indians and thought nothing about it. Him go that way? No sir, never." Jimmy attempted to comfort his friend by placing a blanket under his feet.

Jeff went into convulsions and died at 6:40 a.m. on Friday. Distraught over his friend's death, Jimmy fended off a reporter's question, saying, "My head is just not right today. I thought a good deal of Jeff. It's kinder hard to lose—to lose the kind of friend Jeff was."

The funeral took place at Jack's Cabin on Sunday. Reverend Loder conducted the service, and Mayor Kubler, Melvin Yard and two other men served as pallbearers. Ellen had returned from Crested Butte in time for the funeral, but McLees did not attend because he was confined to his bed due to severe mental strain and heart troubles. The *Gunnison Free Press* said of Jeff's death: "Whatever name the place has had, or whatever character Mickey has borne, he was a friend to those in whom he took any interest, and while much could be said against him, he leaves many to mourn his demise."

On the Tuesday after Jeff's funeral, Ellen put a notice in the *Gunnison News-Democrat* that business would be carried on as usual at Jack's Cabin and that she would be happy to see old friends. A couple of weeks later, the case against Ellen for hiding McLees was dismissed.

In March, Ellen had a grand opening for her ten-pin alley. But running a boardinghouse by herself would not be easy, and she felt more vulnerable as a widow. She filed charges against one of her boarders for borrowing money that he didn't pay back and for running up a large bar tab.

In her autobiography, Ellen recounted many of the events involving Jimmy McLees and Jeff Mickey, but several significant aspects of the way she told her story are at odds with newspaper accounts. The newspaper said that the heavily armed sheriff dispersed the lynch mob at the courthouse, but Ellen says that she dispersed the lynch mob. Another discrepancy was that the newspapers referred to Ellen as "Jeff Mickey's wife," or "Mrs. Jeff Mickey." In her autobiography, Ellen never admits to being married to Jeff Mickey. According to Ellen, Jeff promised to quit drinking if she married him, but after she rejected his marriage proposal, Jeff "told all the boys he did not want to live without me." Ellen's motivation for distortion may have been due to her not wanting to admit she had committed illegal acts, even many years later. If Ellen had married Jeff Mickey without relinquishing her widow's pension, she would have committed fraud, a charge that she would face in a few years.

Jimmy McLees's trial for the murder of Cal Hayzes resulted in a hung jury. His retrial resulted in an acquittal. McLees moved to Montrose, Colorado, a town located sixty-five miles west of Gunnison that was known for its tough characters. A few years later, McLees became upset over the treatment of his son Billy, who had been arrested and put in jail for disorderly conduct by Officer Ladd. Billy witnessed a harrowing event while imprisoned in his cell. A mob broke into the jail, grabbed an accused murderer, hauled the prisoner away and killed him. The vigilance committee wanted to get rid of tough characters that had put Montrose County deeply in debt.

The vigilance committee ordered Billy to leave town, so he wisely heeded the warning by taking the next westbound train. Jimmy should have followed his son's lead, but he did not. Upset by his son being arrested and railroaded out of town, Jimmy arrived in Montrose armed with a double-barreled shotgun and two pistols, remarking, "I don't want to kill anybody, but if he crosses my path [meaning Officer Ladd], I'll let him have it." He drank rum freely and made more threats against the officer. Later that night, Officer Sanborn arrested Jimmy and put him in jail. Around midnight, the vigilance committee disarmed and detained Officer Sanborn, took the keys to the jail, dragged McLees from his jail cell and strung him up over a beam in a stockyard.

2

PROSPECTING FOR GOLD AND SILVER

F or over a year, Captain Jack and John Roberts had kept quiet about the gold deposit on the Ute reservation. In the late summer of 1881, U.S. government troops had used the threat of artillery fire to force the removal of the Utes from Western Colorado onto a reservation in Eastern Utah. In January 1882, prospectors made discoveries in an area near Captain Jack's gold prospect. Excitement ran high with claims of silver that ran $20,000 per ton and expectations that the area would make "hundreds of people millionaires," according to the *Gunnison Daily Review*. The arrival of spring brought many more fortune seekers scouring the hillsides looking for minerals.

Captain Jack and her partners would have to move fast to secure their claim before somebody else found it. In mid-April, Captain Jack, John Roberts and several other men went to the gold deposit in Kezar Gulch, located twenty-eight miles southwest of Gunnison. Captain Jack staked claims called the Big Congo and the Maggie Jack. She sent ore samples to Denver and Boston and received some promising assay reports. She discussed plans for driving a tunnel and erecting a stamp mill. The ore was easy to mill, and there appeared to be plenty of it. She called the place "Jacktown." Despite these early optimistic reports, the excitement soon fizzled, and Ellen did not find any great fortune in the hills of the Kezar Mining District. Even though this initial effort did not pan out at this time, Ellen and plenty of other prospectors would return a little over a decade later to the area south of Gunnison that would become known as the Gunnison Gold Belt.

Captain Jack depicted her adventures as a prospector in photographs taken at her place on the High Drive. *Author's collection.*

Cyrus Wells "Doc" Shores staked claims in the Cebolla Camp area, and he would become well acquainted with Ellen Jack. He was born in Michigan in 1844, and his parents named him after the doctor who delivered him. He obtained his nickname from his brothers, who teased him that he would become a doctor who would ride around on an old horse, dispensing pills to sick people. Doc hoped for a more exciting life than that, and he most certainly would attain it, with his accomplishments inscribed on his gravestone as "Western Colorado's Most Noted Frontiersman, Pioneer, and Lawman." Doc Shores headed west at age twenty-two, and he worked many jobs at places throughout the West: he paid for his passage aboard a steamer by hunting and supplying wild game; he worked as a bullwhacker, hauling supplies to the new gold mining camps in Montana Territory; he provided ties to the Union Pacific Railroad in Wyoming Territory; and he drove cattle from Texas to Kansas. C.W. Shores married Agnes Hoel, and in 1880, the couple headed west to Gunnison, Colorado, where he opened a hardware store. Doc Shores would stake many mining claims in the ensuing years.

Ellen then met her next husband, Redmond Walsh, who had sent men to investigate the gold claims in Kezar Gulch. Ellen referred to him simply as "Walsh." She hoped that he would protect her, because she felt vulnerable as a single woman. Born in Canada, Redmond D. Walsh was fifty-four years old in comparison to Ellen's age of thirty-nine years. Walsh had moved to the United States to work on a railroad in Pennsylvania. He had joined the Union forces and fought at the Battle of Gettysburg. Walsh had been a contractor and superintendent of railroad construction in Ohio and Kansas, and he had taken part in the construction of the transcontinental Union Pacific Railroad.

Redmond Walsh was a partner in the firm of Dunbar and Shafer. The firm had secured contracts from the Denver and Rio Grande Railroad and had finished construction of the rail line for the last twelve miles into Gunnison during the previous summer of 1881. They continued their business in Gunnison by employing 150 men in grading the city streets and digging ditches for the city gas and water pipes. Many of these men boarded at Jack's Cabin.

Walsh was the president of a newly formed branch of the Irish National Land League in Gunnison. In the United States, hundreds of local leagues provided aid to destitute Irish peasants in Ireland and supported their cause of land reform. Walsh's parents were born in Ireland. Redmond was elected as the county delegate for the Ancient Order of Hibernians, an Irish-Catholic fraternal organization.

Redmond participated in a séance held by the Miller brothers, who were described as the "best spiritualistic mediums on this side of the Atlantic." The Miller brothers performed at the courthouse in late June as part of their tour in Colorado. Redmond took part in the slate writing test, in which the Millers asked him to write the name of a departed friend on a slip of paper. Redmond wrote the name of his father and kept the paper in his pocket. A slate was passed into a cabinet, and when it re-emerged, Redmond's father's name, "Richard Walsh," was written on it. The Miller brothers performed other feats typical of such shows, including table-tipping, in which people formed a circle around a table and it mysteriously rose several feet off the floor and spun around several times. The *Gunnison Democrat* questioned whether it was "spooks or jugglers," but no matter how anyone answered the question, most accounts agreed that the Millers put on an entertaining show.

In July 1882, Ellen and Redmond traveled to Denver to get married. During the ceremony, just as the bishop asked Ellen if she would take Walsh as her husband, she heard a man's voice say, "No." The voice seemed close to her, and it caused her to jump back, the ring falling to the floor. Ellen thought that others had heard the voice, too. Despite her misgivings, Ellen E. Mickey, her name as listed on the marriage certificate, went through with the ceremony and married Walsh.

Ellen wrote this troubled description of her marriage ceremony in her autobiography over two decades afterward. In the intervening years, her memory may have been negatively distorted. When a marriage goes downhill, people remember their past as a couple in negative ways. They tend to remember their early years as chaotic, and they have trouble remembering anything positive about their spouse.

Ellen had hoped that Walsh would protect her, but he soon became her adversary. Ellen was persuaded by Walsh to loan money to Dunbar and Shafer so they could pay their workers. Ellen did not get her money back, so she filed suit against them in January 1883 for $1,107 in unpaid bills. Ellen won the case, but she also incurred what she considered to be outrageous legal fees that she didn't fully pay, and her unpaid bills would haunt her over the next several years.

Ellen continued with her mining ventures by locating two fire clay deposits. The *Gunnison Daily Review* described, "Mrs. Jack has jumped Donnelly and Harper's kaolin placer claim, two miles east of town. There is likely to be some fun before the matter is settled." The newspaper referred to her as Mrs. Jack, even though she was married to Redmond Walsh at this time. A week later, Mrs. Jack, "with a revolver, ran two workmen off the fire clay discovery two miles east of town." Despite whatever disagreement that may have existed between the parties, Ellen and Redmond sold the Walsh Lode and the Lewis Lode to John Donnelly and Samuel W. Harper in May 1883.

According to Harper, Redmond was taking advantage of Ellen. The claims belonged to Ellen, but the two buyers gave Redmond $200 in cash, and they gave Ellen a note for the remaining $300. Redmond later got the note from Ellen and cashed it at the Bank of Gunnison. Redmond had told Harper that he was acting in his wife's interest.

As a result of this case, as well as the Dunbar and Shafer case, the cashier at the Bank of Gunnison warned Ellen about Walsh: "You are a Mason's widow, and I am a brother and am sorry to have to tell you that several of us have been watching your affairs a little, and we find out that you have been duped by the man Walsh. He is not what he represents himself to be." He also warned Ellen to "be on your lookout for that man; he would not hesitate to take your life."

Ellen soon heard with her own ears about Walsh's dastardly plans. Ellen also found out that Walsh was a bigamist. He had been married to another woman when he married Ellen. At Jack's Cabin, she heard Walsh arguing with another man over money. A mysterious voice said, "Come in and be silent." She hid in a nearby room and listened to the argument.

The man said to Walsh, "If she gets onto your having a wife living, she will put you through for what you have done."

Walsh said, "I will take care that she does not get me, for I will put her where she will not do any harm."

"I want the money you promised me, and if you do not get it for me, I will send to Leadville for my sister to come here and tell the captain that she is your lawful wife, for I am determined that you shall not dupe me with your damned lies."

Ellen met with Walsh's other wife, Anna Gandley, and Ellen was convinced that she was in fact his wife. According to Ellen, Walsh drank heavily, and he was intoxicated on a regular basis. Ellen filed for divorce in the Gunnison County Court on June 23, 1883, less than a year after their marriage. The court denied her case.

Five months later, Redmond and Ellen agreed to form a five-year partnership for mining purposes. Redmond gave Ellen a one-half interest in his two mines in Pitkin County, and she gave him a one-half interest in her two mines, the Big Congo Mine and the Jacktown Mine. They both agreed to give each other a one-half interest in any mine that either one of them later acquired. Ellen said that she did not sign this agreement that Walsh presented to her. After she threw it into the fire, Walsh grabbed her and thrust her head toward the flames, burning her face.

$$\sim\!\!\sim\!\!\sim$$

ELLEN JACK NEXT PURSUED hidden treasures by owning the Black Queen Mine. She bought a half interest in the mine from L. Virginia Ditto and Wm. Strite Ditto for $500 in February 1884. But the path to riches would not be an easy one, and she would have to fight many people along the way, some of whom she may have viewed as trusted friends.

The Black Queen Mine was in the remote mountainous area thirty-five miles north of Gunnison, on Sheep Mountain, near the town of Crystal. The area had been explored by prospectors in the early 1870s, but it wasn't until 1880 that about one hundred prospectors occupied camps along Rock Creek. A year later, the twenty-six-year-old Al Johnson and his younger brother Fred opened a store in Crystal. Northwest of the mining camp, over five hundred men worked eighty-four claims on Sheep Mountain.

Captain Jack co-owned the Black Queen Mine with two other men, who had each bought a quarter interest in the mine from one of the original

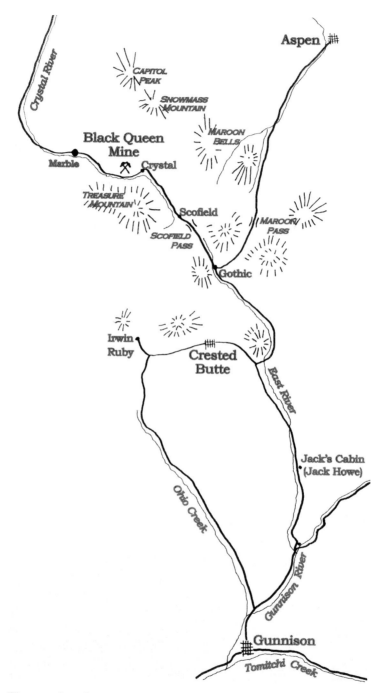

The route from Gunnison to the Black Queen Mine. *Drawn by the author.*

Above: Snow bridge over Rock Creek (later renamed the Crystal River), showing the difficulty of travel. *Author's collection.*

Opposite: Prospector's cabin near Crystal. *Author's collection.*

claim owners. George Farnham, age thirty-one, and Lark Young, age twenty-nine, had previously lived in Iowa. George's wife had died, and he left his daughter with his wife's family when he went to Colorado. Lark Young had lived in the Colorado Springs area in 1880 with his wife, daughter and two brothers before they all moved to Gunnison.

Captain Jack bought into the mine at just the right time. In the previous two years, George Farnham had sunk a twenty-foot incline shaft with hand tools, and he had hauled ore and bailed water out of the mine. The development work began to pay off, and the Black Queen Mine started to receive acclaim. In July 1884, the *Rocky Mountain News* described the Black Queen Mine as "a mineral wonder." Assays of the ore showed values ranging from $87 to $587 per ton. By August, even richer ground had been reached, producing ore that assayed at 9,300 ounces of silver per ton, with chunks of ore as big as a man's head. Burros loaded with ore trod down the steep two-mile-long trail to Crystal, slogged up the treacherous wagon road over the 10,707-foot Scofield Pass and delivered the payload to Colonel Rose's sampler in Crested Butte. The sampler took in ore from many small mines, concentrated or smelted the ore and then shipped and sold it to a larger smelter.

Captain Jack, Farnham and Young received an offer of $100,000 from unnamed Pennsylvania investors for the mine, but they held out for an

Crystal in 1906. In the center of the photograph, the scar along the side of the mountain is the route to Scofield Pass. *Author's collection.*

asking price of $125,000. Not anxious to part with their mine when one shot of explosives in it could produce thirty sacks of almost pure silver, the owners optimistically anticipated earning at least $100,000 in a short time if they held onto their mine. The deal fell through.

Captain Jack and her partners leased the Black Queen to three Welsh miners, William Parry, Griffiths D. Griffiths and Evan Jones, from September 1884 to July 1885. The three men co-owned and worked numerous claims in the area. The lessees agreed to sink a shaft to a depth of 125 feet or more. They would keep 70 percent of the gross production, with the other 30 percent going to the owners. Due to the steepness of the terrain, they excavated a flat area out of the mountainside and created a shelf on which they built a cabin, shaft house and ore house. The miners benefitted from a mild start to the winter, which kept the road passable and allowed them to ship one thousand pounds of ore in December to the Moffat smelter in Gunnison. By mid-January, storms had dumped enough snow to make travel impossible, so the miners had to store the ore before they could ship it again in the late spring.

It is not clear how much profit Captain Jack and her partners were making from the operation at this early stage. Estimates of the value of the ore taken from the mine varied widely. George Farnham stated that he shipped twenty tons of ore that had a value of $1,000, but he had sunk at least $4,000 into development of the mine. The *Rocky Mountain News* listed tonnages and values for several of the leading mines in the area for 1884, reporting shipments of eighty tons from the Black Queen, with a value of $4,000.

Crystal residents enviously watched the processions of burros loaded with ore passing through their fledgling camp. According to George Farnham, Albert Johnson and his partners "came up there [to the Black Queen] after we had struck mineral, and there was quite an excitement to see if their ground [the adjoining Excelsior claim] did not take in the discovery and found that it did not, and when they went down, they took some of the samples of the ore from the mine with them." Johnson spread the news of the find to the newspapermen in the nearby town of Gothic by showing them a fist-sized piece of nearly pure silver ore. Al Johnson did not share Farnham's opinion that the mineral discovery was on the Black Queen claim. In February 1885, Johnson vied for the valuable silver ore by declaring that the mineral vein was on his adjoining Excelsior property.

In 1881, Charles Cheese first staked the Excelsior claim. He had already found the Black Queen vein nearby. Based on assays of the mineral samples, he thought that the Excelsior was the more valuable property, so he did not do any further work on the Black Queen vein. In surveying the Excelsior claim on July 28, Charles Cheese directed U.S. deputy mineral surveyor Robert Sterling to draw the south boundary of the Excelsior claim above the Black Queen workings to avoid any conflict with the adjoining claim.

L. Virginia Ditto, Wm. Strite Ditto and William Langman claimed the Black Queen Lode one day later, on July 29. The *Gothic Miner* had cheerfully reported that everyone was working their own claims, with no reports of claim jumping or expensive litigation suits, but they warned that this situation could soon change. By the end of the summer, with one thousand prospectors blanketing the area, each seeking a fortune, disputes arose over competing mining claims. A miner's union was formed to settle disputes without traveling the long and arduous distance to Gunnison. The newspaper recommended that the miner's union should hang a few claim jumpers to take care of the problem. But they did not have to take such drastic measures. The miner's union settled a dispute that Langman had with adjoining claimholders. With that issue settled, Langman sold his portion of the Black Queen Mine to George Farnham and C.W. Young.

Now, Johnson staked his own claim on the Excelsior boundaries, and he boldly maintained that he owned the Black Queen's ore. This conflict would simmer for a couple of years. When Captain Jack and her co-owners received good offers to sell their mine, Johnson would reassert his argument that the mineral vein belonged to him as the owner of the Excelsior claim.

The well-known author Helen Hunt Jackson traveled to the Gunnison Country in 1883, and she described how prospector Jim Brennan had named two adjoining drainages in the Elk Mountains, near Crested Butte. She wrote, "When Jim Brennan named these basins and gulches, nothing was farther from his mind, probably, than the idea of speaking in parables. But if he had so meant, he could not have done better. Poverty Gulch and O-Be-Joyful Creek, the two will be found always, side by side, as they are in Gunnison County. Only a narrow divide separates them, and the man who spends his life seeking gold and silver is as likely to climb the wrong side as the right."

3

ASPEN

Ellen suffered severe pains from a gallstone attack, and a woman whom she referred to as "Aunt Susan" took care of her during this illness. After Ellen recovered, Susan Bryan and an unnamed man helped Ellen pack her belongings into three horse-drawn wagons. The journey to Aspen took four days. They most likely took the sixty-mile route north from Gunnison, over the 11,800-foot Maroon Pass and down the drainage of East Maroon Creek into the bustling silver mining town.

Susan Bryan, described as Black or Mulatto on census records, was born into slavery in 1834 in Missouri. In 1881, Susan Bryan came to Gunnison by team and wagon, accompanying Jesse and Alice Corum and their two children. Alice Bryan Corum's family had owned seventeen enslaved people in 1860, one of whom was most likely Susan Bryan. Mrs. Bryan was loyal to Jesse Corum, who had fought for the Union in the Civil War. In Gunnison, the Corum family lived on East Tomichi Avenue, near Jack's Cabin. Mrs. Bryan's son, Richard Bryan, arrived a year later, and he lived in a nearby household with his wife, Hester, and his sister. Richard built a cabin for his mother that she would live in for the rest of her life. Susan Bryan would become a well-known figure in Gunnison. A cheerful person, she would be known for taking care of many people throughout her life.

The term *aunt* was sometimes used when referring to Black women due to Jim Crow etiquette, in which white people might not accord Black people a title of respect, such as "Mrs." or "Miss." Ellen Jack referred to Susan Bryan several times as "Aunt Susan." The Gunnison newspapers referred to

Ellen Jack traveled over Maroon Pass on her way to Aspen. *Author's collection.*

her as "Aunt Susan," and everyone in town knew her by that name, but the newspapers also referred to her as Mrs. Susan Bryan in some issues.

Ellen had hoped to give Redmond the slip by moving to Aspen, but he followed her and caused more trouble. Ellen filed for divorce again, this time in Pitkin County, in November 1884. She accused Redmond of extreme cruelty. Around this time in the western United States, 20 percent of divorces were granted based on charges of cruelty. In December, Redmond filed a suit against Ellen for his quarter interest in the Black Queen Mine. He asserted that, as her mining partner, he had a right to half of her properties.

Ellen accused Redmond of being habitually drunk. Attitudes about whether a woman should endure the abuse of a drunken husband varied widely. The *Gunnison Review-Press* opined that women should heed the decision of the Iowa Supreme Court, which denied a woman's divorce from her drunkard husband because she knew he was a drunk when she married him. She should put up with her fate of living as the wife of a drunk.

Other people disagreed with that view. The Colorado Humane Society formed in Denver in 1881 with the prevention of animal cruelty as its first objective, followed by its second objective, the "prevention of cruelty and neglect of children and protection of the wives of intemperate husbands." A report of one month's activities showed eight cases of cruelty to women,

Aspen. *Author's collection.*

thirteen cases of cruelty to children and twenty-five cases of cruelty to animals. The most common outcome was that the offender was admonished, but in a few cases, the offender was arrested and fined.

Also in December 1884, Ellen leased a building for business purposes from Mrs. Sarah Adair. John and Sarah Adair were early pioneers in Aspen. John Adair discovered what would become one of the best mines in Aspen, the Mollie Gibson; he became the superintendent of the Josephine Mining Company; and he was elected as a trustee in the newly formed city government. He died in 1881 from injuries sustained in an unfortunate accident in which a wagon ran over him. Many friends attended the funeral of this well-respected man.

In the building leased from Mrs. Adair, Ellen opened a saloon and boardinghouse that she said brought in $400 a month. Saloonkeeper Frank Royer and miner James Murphy, along with two other men, lived at this same address. Frank Royer had known Redmond Walsh and John Young, the brother of Black Queen co-owner C.W. Young, in Colorado City. Royer had been a police officer in Pueblo, and he distinguished himself in that job. For example, he rushed to the rescue of a woman who had lost control of her team of horses and buggy. The *Pueblo Daily Chieftain* described Frank as a "gallant young man…always ready to give aid to fair females in distress." Royer had pursued mining ventures by working a coal seam near Pueblo.

Toward the end of January 1885, less than two months into the ten-month agreement, Mrs. Adair filed suit against Ellen to regain possession of the building. Ellen lost the initial suit, but she filed an appeal. Mrs. Adair sent Ellen a notice to vacate the property. Ellen refused to leave. Mrs. Adair filed a complaint in the Pitkin County Court, alleging that Ellen had subleased the property to people who were engaged in acts that were against the law in Colorado. Once again, Ellen refused to move out. Ellen was preoccupied with her divorce trial in April, and she answered Adair's charges in May.

In her autobiography, Ellen Jack recounted an incident in which Walsh tried to kill her. She awoke one night to the sound of two men outside of her bedroom window: one was Walsh, and the other was a morphine fiend. She got up, saw the men, fired her gun at them and wounded Walsh in the arm and the leg. She discovered dynamite that Walsh had placed there to blow her up. Ellen learned from a friend that Walsh would try again to kill her to gain possession of the Black Queen Mine.

According to witness Jim Coles, who was in his livery store across the street from Captain Jack's boardinghouse in Aspen, a bright flash lit up the night sky, followed by a booming explosion. The miners who had been sleeping in the Vallejo boardinghouse spilled out of the building, and other excited onlookers came to investigate the cause of the commotion. A five-foot-long fuse had ignited a stick of powder that someone had thrown onto the roof of the boardinghouse. The rags wrapping the powder had not caught fire; otherwise, the incendiary device would have done more damage. The *Aspen Daily Times* asked, "Why did it occur?" But it provided no answer. Captain Jack was not mentioned in the newspaper story, but her boardinghouse was located at the corner of Cooper Avenue and Hunter Street, where the incident occurred. It is likely that Ellen embellished the part about shooting Walsh. There is no evidence in the Aspen newspapers of her shooting Walsh, an event that most likely would have been reported.

Ellen felt remorseful for marrying Walsh, and she asked herself, "Why did I marry him, and what is this power force?" She explained her fate by referring to Rosicrucian beliefs:

> *I was conscious that I was doing wrong, and even the ring was dashed to the floor by unseen hands. It could not be spirit power but a power far stronger than that which forces us to our destiny, and we ought to be on our guard all the time for strangers that we know not, for some people carry a straight light around them that is destruction to one that carries the opposite light.…There is a triangle of three powers that govern both heaven and earth—electricity, vibration, and this force power. I know not what to call it, but it comes in waves and in different colors and does its work according to its color. It is plain to see that Walsh attracted the dark, destructive way and came to a house of light and brought nothing but loss and misery, and would have murdered me if the same power only in the light was the strongest around me, so failed the dark. That is what saved me.*

The divorce case of *Ellen E. Walsh v. Redmond D. Walsh* took place in April 1885 in the Pitkin County Courthouse on Cooper Avenue. Newspapers would comment on the sensational testimony in this case in the years following the verdict. The *Aspen Daily Times* wrote that "the charges and counter charges made during the trial were extremely disgusting."

A jury of six men, with Judge Thomas A. Rucker presiding, would determine whether Ellen had grounds for divorce. The case drew a "crowd of hungry curiosity seekers," who packed the courtroom to listen with "undivided attention" to the "salacious" details of the case. Ellen's attorneys made a request to exclude the spectators from the courtroom, and the court agreed.

Ellen charged that her husband had "struck and kicked and beat her in a violent and brutal manner." He verbally abused her, calling her a "whore and a bitch," and he threatened to kill her. He was habitually drunk after February 1, 1883. She explicitly described his sexual abuse. Redmond Walsh's lawyer moved to strike this last bit of testimony, but the motion was overruled.

Elisabeth Hostrawser corroborated Ellen's claims of cruelty. Elisabeth testified in a written affidavit that she lived across the street from Redmond and Ellen Walsh in Gunnison in 1883. One morning, Ellen came to her residence and said that Redmond had beaten her and had attempted to put her head in the oven. Ellen showed Elisabeth a bruise on one of her limbs that had resulted from Redmond kicking her. Ellen stayed there for the rest of the day and overnight, and she went to Leadville the next day.

Redmond Walsh denied all charges; he said he had not committed acts of cruelty toward his wife, he had not called her names, he had not threatened to kill her and he had not been habitually drunk. He asked for a dismissal of her complaint. He charged Ellen with committing adultery with Frank Royer. The *Aspen Times* printed the "secret" that Redmond was stubbornly defending himself in the divorce suit because he wanted a part of the valuable Black Queen Mine.

Judge Rucker issued extensive instructions to the jury. The jury had to judge the credibility of the witnesses. If the jury believed that Ellen Walsh had committed adultery, they should rule in favor of Redmond Walsh. The burden of proof was on the plaintiff, Ellen Walsh. The judge instructed the jury that "a single beating, bruising, striking, or whipping of a wife by her husband is sufficient to establish a charge of extreme cruelty. And that if the jury believes from the evidence that the defendant [Redmond] is guilty of striking or beating in any manner the plaintiff [Ellen] one or more times that they may find for the plaintiff." The instructions also defined habitual drunkenness as "that degree of intoxication from the use of intoxicating drinks which disqualifies the person for a great portion of the time from properly attending to business or which would reasonably inflict a course of great mental anguish upon an innocent party."

The jury found in favor of Ellen Walsh. After the divorce, she resumed using her previous name, Ellen E. Jack.

While the lurid details of the case attracted the prurient interest of curious spectators, the Aspen press harshly denounced it. The *Aspen Daily Times* decried that the case "brought out more repulsive and disgusting details than any other case that ever was tried in Pitkin County....Both parties to the case are advanced in years, which fact tends to increase the repulsive character of the whole proceedings." The description "advanced in years" referred to Ellen's age of forty-two years and Redmond's age of fifty-seven years. The newspaper suggested that the "county courtroom should be fumigated after the Walsh divorce suit."

Now that the divorce case was over, Ellen faced the charges from Sarah Adair. In May, Mrs. Adair more explicitly described why she was trying to evict Ellen from her property. She charged that the people who subleased the property were engaged in "acts of open lewdness and notorious acts of public indecency, tending to debauch the public morals." Adair charged that these acts went on with Ellen's full knowledge and consent and that she had kept a "lewd house…for the practice of fornication and prostitution" that led to the "encouragement of idleness, gaming, drinking, fornication, and public misbehavior." She alleged that these acts were continuing to such an extent as to be a "matter of general and public notoriety."

Ellen answered this complaint by making somewhat contradictory statements: that the sublessees were not breaking any laws and that Sarah Adair knew the businesses and employment of each sublessee. Ellen also denied operating a lewd house but then stated that "said acts, if any, were with the full knowledge and consent" of Sarah Adair. In reply, Mrs. Adair denied any such knowledge. It was a common strategy of property owners to deny awareness of the type of activities that transpired in such places. Ellen settled the dispute by purchasing the properties from Sarah Adair for $1,600 in June, and the court dismissed the legal suit.

Was Ellen Jack running a house of prostitution, as Sarah Adair had charged? Ellen owned two town lots separated by an alley, with one lot facing the respectable part of town on Hyman Avenue and the other lot facing the saloon district on Cooper Avenue. In February 1885, during the

One of Ellen's properties faced Hyman Avenue. *Author's collection.*

time of Mrs. Adair's accusations against Ellen, the town council passed a resolution prohibiting prostitution north of the alley separating Hyman and Cooper Avenues. Ellen's lot facing Cooper Avenue contained a saloon, an icehouse and two smaller buildings. Six other saloons lined that one-block stretch of Cooper Avenue. It is possible that the charges against Ellen were true, given that the saloon on her property was in the area where prostitution was allowed.

Located one block south of Cooper Avenue, Durant Avenue contained most of Aspen's brothels. Town ordinances passed in 1881 prohibited prostitution and set a penalty between ten and one hundred dollars for conviction. Even though city and state law prohibited prostitution, in practice, the marshal collected a five-dollar fine per month from each prostitute, thus establishing what amounted to a license. Marshals often skimmed off a portion of the proceeds for themselves, and city coffers depended on this income. The fines also kept the prostitutes poor.

In contrast to the glamourized version of prostitution often seen in movies, television shows and tourist attractions, scholars have focused on the reality that most prostitutes led lives of chaos and destruction. Aspen newspapers often mocked prostitutes. While a fire burned in one building on Durant Avenue in the middle of the night, many "ladies and gentlemen" lined the street, and "the shadowgraphs presented after the interior had been lighted were thoroughly appreciated by the spectators." The women also suffered from assaults. In one such incident, Julia Smith, described as "one of those handpainted nymphs who preside on recreation row," endured a disfiguring attack from a "fiend" who threw acid on her face. The frequently used word *inmate*, in reference to a prostitute, comes closest to describing prostitutes' actual life circumstances. Due to the high rate of drug addiction, it was sometimes difficult to tell whether a death by suicide was intentional or not, as was the case with Jennie St. Clair, who died of an overdose of morphine.

Ellen Jack's operation of a saloon—and quite likely a brothel—put her on the margins of respectable social circles. The legal system enacted laws and fines to constrain prostitutes from escaping their situation. Similarly, the legal system would allow Captain Jack's lawyers to impose and collect excessive fees for their services, which would constrict her ability to own and profit from the Black Queen Mine and live her life as she pleased. Captain Jack's view of herself as having a high status because she was a widow of a naval officer carried little weight in her current situation. Her ownership of a working-class saloon and boardinghouse placed her in a class of people that was viewed by the community's leaders as beneath them.

Ellen Jack's legal troubles with Redmond Walsh continued when he sued her for a part interest in the Black Queen mine. Judge Gerry heard the case in the district court in Gunnison. Redmond Walsh introduced into evidence a contract that Ellen had signed, which gave him a one-fourth interest in the mine. Ellen testified that the document was a forgery. Several men testified on Ellen's behalf, including Frank Royer and James Murphy, as well as two of Gunnison's bankers, Sam Gill and Eugene P. Shove. Judge Gerry ruled in favor of Ellen, as Mr. Walsh had not produced convincing evidence of his claim. Ellen kept her half interest in the Black Queen Mine.

Redmond Walsh testified against Ellen in a case brought by James Cox in the fall of 1885. James Cox charged Ellen with owing him $500 plus interest. Cox had worked as a miner at the Doctor Mine, located north of Gunnison. He received his pay, came into Gunnison and stayed at Jack's Cabin in March 1884. According to the testimony of Redmond Walsh and James Cox, James had gone out to buy a pair of shoes, and when he returned to the barroom of Jack's Cabin, Mrs. Walsh had arranged a scheme. He gave her $500 in exchange for a promissory note that was secured by the mortgage to Jack's Cabin. Cox claimed that he was temporarily insane during this transaction because he was intoxicated; therefore, he could not be held responsible for entering any legal contract.

About a week later, James Cox wanted to get some money to travel to Salida, so he approached Redmond, who was behind the counter in the barroom of Jack's Cabin. Ellen stood next to James at the counter, insisting that he should give her the promissory note for safekeeping, because he was likely to lose it. He did not want to give it to her, and he kept it tucked away in his pocketbook. She made a bet that he didn't have it. To prove her wrong, he took out the note and showed it to her. Ellen snatched the note from his hand. James Cox testified that Mrs. Walsh "made a dive and took it out and kept it. I didn't care for having all the people of the city around me, or I would have tackled her for it." James never saw the note again, and he accused Ellen of destroying it. Additional damaging testimony against Ellen came from four men, who stated that her "general reputation for truth and veracity was bad."

In contrast to James Cox's story about Ellen scheming to take his money, Ellen's story was much different, and it made her seem like a motherly figure who was taking care of an irresponsible child. She testified that James Cox was worried about keeping his money in a safe place. While staying at Jack's Cabin, he had fallen asleep and left his money bag under a table. When he woke up, he was upset and worried that he had lost all his money over a fuss with some woman at a saloon. Ellen arranged for the transfer of the mortgage to Jack's Cabin in exchange for the $500. She agreed that she had given James Cox a promissory note for $500. However, she claimed charges of $300 against him for money she had given back to him and for his room and board.

This time, the verdict went against Ellen. The jury awarded James Cox $672. Ellen appealed the case to the district court and to the Colorado Supreme Court, but she lost both cases. Ellen Jack did not mention this court case in her autobiography.

James Cox was represented by John Kinkaid and Sprigg Shackleford. Kinkaid, age twenty-eight, had the "nobbiest law office in Gunnison." Ellen was represented by Alexander Gullett and Henry Karr. Gullett had served in the Union army. He was appointed clerk of the Committee on Indian Affairs of the House of Representatives in Washington, D.C., where he became friends with Senator Henry Teller of Colorado. All these lawyers would play prominent roles in Ellen's future legal battles.

Ellen Jack retained control of the Black Queen Mine after these clashes with Redmond were over. Shortly before her divorce case in April, Ellen sold her Gunnison properties and the Black Queen Mine to Edward J. Stewart for $2,000. Ellen said that Stewart was Frank Royer's brother, who was involved in mining in Leadville. They had an agreement that the money was a loan, with the understanding that the properties would be hers again if she paid him back. Ellen said that she made the agreement to "save any litigation," which probably meant that she was selling the mine so that Redmond Walsh could not possess it if he won his cases against her. After legal troubles with Redmond subsided, E.J. Stewart sold those properties back to Ellen in December for the same amount of money. Questions would remain about whether there was any person named E.J. Stewart, or if Frank Royer, who sometimes used the last name Stewart, may have assisted Ellen in making sure she retained ownership of the mine. But Ellen would contend with Redmond one last time when he testified against her for pension fraud.

4

ELLEN JACK FIGHTS TO KEEP THE BLACK QUEEN MINE

By the fall of 1885, the Black Queen had attained prominence as a productive mine. Frank Royer leased the mine, and he hired four other men. They bailed out the water, sunk the incline shaft to a depth of 130 feet, worked two levels in the mine and shipped ore. In November, the good news continued when Ellen reported that her crew of eighteen miners had made a big strike that ran over 1,600 ounces of silver per ton.

The brutal winter of 1885–86 saw numerous heavy snowstorms. Captain Jack did not let difficult weather conditions stop her from taking one mid-December trip. She traveled on horseback for seven miles from Crested Butte to Gothic during a snowstorm. The next day, she set out to travel the twenty-one miles to Aspen, but she turned back after several men convinced her that the route via the Maroon road was impassable. With the temperature at eight degrees Fahrenheit below zero the following day, she left Gothic and made her way into Aspen.

In Crystal, Al Johnson delighted the miners with a special Christmas Day feast of "oyster stew, beef's heart, potatoes a la crème, roast turkey with oyster dressing, cranberry sauce, fricasseed chicken, corn and tomatoes, rolls, salad, French pickles, coconut pie, lemon pie, cakes, nuts, candies, tea and coffee," thus departing from the typical miner's fare of biscuits, bacon and dried fruit. A particularly severe January snowstorm dropped four feet of snow in Crystal. A snow slide ran down the nearby White House Mountain and swept away two miners and their cabin. In February, James Murphy, who had been working on the Black Queen Mine, was traveling to Aspen

Crested Butte. *Author's collection.*

when he became exhausted and sank into the snow. With death close at hand, Robert Sterling and another person from Scofield rescued Murphy and revived him with brandy.

Hardy souls who wintered over in the Gunnison high country found a way to play in the deep snow. Robert Sterling and Al Johnson both served as officers in the Gunnison County Snowshoe Club (skis were called snowshoes back then). The club held a major racing event in Crested Butte in February 1886. A crowd of around one thousand spectators traveled from Gunnison and the nearby mining camps to watch the race. In the final round of competition, four skiers sped along the steep downhill route, with Al Johnson coming in second place a mere two feet behind the winning racer.

Crystal residents formed a very tight-knit community that would support Al Johnson in his attempt to gain control of the Black Queen Mine. Johnson kept busy scheming for a way to possess the riches of the Black Queen. He would have to make his move soon, because Captain Jack and her co-owners were receiving valuable offers for their mine.

The Black Queen Mine attracted more interest from potential buyers in early January 1886. Frank Royer met with the U.S. senator from Colorado George Chilcott and others in Gothic, and from there, they traveled to Crystal to view the mine. The asking price was $80,000. In the previous few years, development of the mines near Crystal had proceeded at a slow pace, but with the influx of much-needed capital, prosperity could be close at hand.

The possibility of a rich mine attracted the greedy fingers of everyone who thought they could grab a piece of it. On February 16, 1886, Byron

Captain Jack looking for a company to buy the mine. *Author's collection.*

Shear and Isaac Johnson filed a suit to force a sale of Ellen's mine. These prominent attorneys had represented Ellen in her court case against Sarah Adair, and they had provided other legal services. They claimed that she had paid them only $90 in legal fees and that she still owed a balance of $863. Shear and Johnson obtained writs of attachment against the Black Queen Mine and Ellen's property in Aspen as payment for their debt if they won the case.

According to Ellen, the attorneys charged her "the most outrageous prices....Such robbery I never thought would be tolerated, for there was not $100 due them." The fees amounted to double the average yearly wages of a working man. These exorbitant fees and collection procedures could be considered another form of "courtroom mining." The term typically refers to instances in which competing claim owners fought court battles over possession of mining claims. Many famous stories in mining history involved big cases with wealthy, powerful litigants on both sides. Less well known are cases in which lawyers charged excessive fees and used their social positions and connections to force the sale of a mine owned by someone of more limited means.

The high lawyers' fees were not unusual. In the California gold rush, lawyers realized that a lot of money could be made from "mining the

The miner's dream of finding a rich mine. *Author's collection.*

miners." Some early California mining districts banned lawyers, because miners viewed them as "scavengers and bloodsuckers." But disputes arose among miners over a variety of issues, and lawyers charged high sums for settling those disagreements in the legal arena. One lawyer in San Francisco made $3,300 in ten days, and he wrote to his parents about his experience: "You have no idea of the extent of litigation in this city or the size of the fees."

In Nevada, an estimated 20 percent of the $45 million wealth of the Comstock Lode ended up in lawyers' pockets. William H. Stewart made his fortune as a Comstock lawyer, earning $500,000 in four years as a "servant of power," as one of his biographers called him. He was one of the principal U.S. senators who developed the Mining Laws of 1866 and 1872. The laws standardized certain procedures for staking a claim: the size of the claim, proper recording procedures, the $100 worth of work per year necessary to maintain the claim and the process of patenting a claim. The laws also described the apex rule, which resulted in much litigation between adjoining claim owners. The Mining Law drew immediate criticism, with one critic calling the law "an act to encourage litigation and for the benefit of lawyers and not to promote the real interest of the miners or increase the product of the mines."

Ellen resorted to her own shenanigans to try to hang onto her mine. She sold her half interest in the mine to her brother, William Elliott, on February 17, the day after Shear and Johnson filed suit against her. Because her brother lived out of state, Ellen would be his attorney in fact, which meant that she had the legal authority to make transactions for him. Several lawyers would accuse her of selling the mine to her brother to avoid paying off her debts.

Ellen lost the case brought by Shear and Johnson. The Pitkin County Court ordered the sale of the Black Queen Mine to satisfy the debt. Acting as her brother's attorney, Ellen would try to figure out a way to hold onto the mine.

Ellen also owed money to Alexander Gullett and Henry Karr for their work over the previous years on several cases, including the Dunbar and Shafer case, her divorce case and the James Cox case. Ellen accused them of charging excessive fees and of being negligent in losing the James Cox case. Gullett and Karr won a court judgment against her for $1,422 in fees.

The reports from the Black Queen over the summer showed that there was something worth fighting for. The nationally well-known *Engineering and Mining Journal* described the Black Queen Mine as one of the most prominent mines in the Sheep Mountain District. The mine had produced some good ore, and parties from Denver were negotiating to buy it, but the owners had placed a high price tag on their property.

Aware of recent valuable discoveries from the Black Queen, Al Johnson continued with his attempt to gain possession of the mine. He instructed U.S. deputy mineral surveyor Robert Sterling to survey the boundaries of his adjoining Excelsior claim. Anxious to show that his claim extended onto ground claimed by the Black Queen, Johnson instructed Sterling to make sure that the survey would show such a result. In mid-July, Albert Johnson ousted the owners of the Black Queen Mine from their claim. Johnson asserted that a portion of his mining claim overlapped with the Black Queen. The discovery shaft of the Black Queen lay just outside what he asserted was the boundary of the Excelsior Lode, but the incline shaft went under the sideline of the Excelsior claim. Johnson argued that the ore coming out of the Black Queen came from the Excelsior. Johnson filed for a patent on the Excelsior Lode on July 26.

Johnson began publishing the newspaper the *Crystal River Current* in September 1886, in addition to carrying out his duties as postmaster and storekeeper. The name of the paper referred to the recently renamed Rock

Al Johnson's store (*building on the left*) and the Crystal Saloon (*middle building*). The trail to the Black Queen Mine goes up through the aspens. *Jane Bardal, photographer.*

Creek, now called the Crystal River. He bought the office of the *Crested Butte Gazette* and moved the outfit to Crystal. As an energetic and tireless booster of the mines surrounding Crystal, Al also delivered the latest mining news for publication in Aspen and Gunnison newspapers. Through these newspapers, he could broadcast his version of events.

Al Johnson, along with other "Rock Creek boys," joined with citizens of Crested Butte to hold a farewell party for U.S. deputy mineral surveyor Bob Sterling in early October. A band played, men made speeches and the whiskey flowed freely. The next day, Sterling departed to see his parents in his hometown of Cincinnati. After that, he planned to spend two years in England, where he had previously attended the University at London.

Ellen Jack and her co-owners challenged Johnson's attempt to take over a portion of the Black Queen claim. They charged Johnson with trying to cheat and defraud them out of their mine. In reaction to Al Johnson's filing of a patent application for the Excelsior, Ellen Jack, William Elliott, George Farnham and C.W. Young filed a protest and adverse claim in the United States Land Office in Gunnison in September 1886. An adverse claim may be filed if other persons have competing claims to the same land that someone else is trying to patent. They asserted their right to the Black Queen claim and the area of dispute with the Excelsior. Ellen Jack and her mining partners had succeeded in developing a paying mine, but now they would have to fight to retain ownership of their bonanza.

Mining historian Otis E. Young Jr. called this type of litigation a "legalized form of thievery."

> *The game lay in locating, or claiming, the lode where it cropped out of the ground (at its apex), since he who held the apex held all. This was a simple and logical rule, provided that mineral veins were always continuous and discrete—which, in fact, they seldom were. Because of this geologic difficulty and because only the Almighty knew for certain the exact extent and ramifications of an undeveloped lode, such uninspired reasoning soon revived an ancient and sometimes highly profitable occupation—courtroom mining, or as it was more formally called, apex litigation.*

In many cases, "the facts were so hard to determine that sheer legal showmanship made a great deal of difference in the judgments and the awards." And many lawyers charged exorbitant fees for their services.

Meanwhile, the lawyers tried to force a sale of Ellen's half interest in the Black Queen to satisfy the judgments against her. Sheriff C.W. Shores

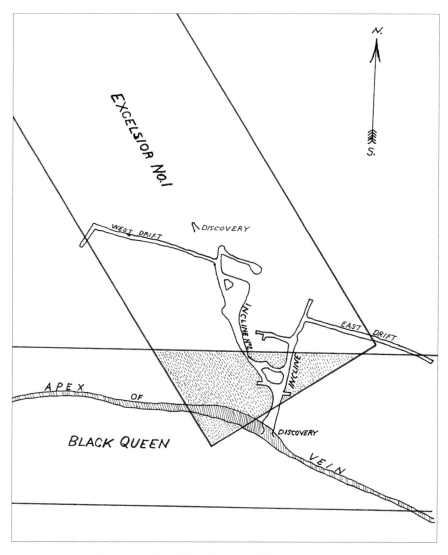

A contested area of overlap of the Black Queen and Excelsior mining claims. *Drawn by the author.*

scheduled a sale for December 4, 1886. C.W. "Doc" Shores had been elected sheriff of Gunnison in the fall of 1883, and he would hold the position until 1892. He wrote his memoir toward the end of his life, recounting his many exploits in capturing murderers and train robbers. He did not mention the more mundane aspects of his job, such as selling property to satisfy debts, nor did he mention any of his mining claims. His name would become attached to several of the lawsuits against Ellen, probably because he supplemented his meager sheriff's salary by collecting debts. He would try to sell Captain Jack's mine several times before she finally figured a way out of her predicaments.

Before Doc Shores could sell the Black Queen, William Elliott went on the offensive against the lawyers who were trying to force a sale of the mine. William Elliott filed suits against both groups of lawyers: Shear, Johnson and Shores; and Gullett and Karr. He argued that on February 17, 1886, Ellen Jack had sold her half interest in the Black Queen to him. He owned the mine, so it could not be sold to pay her debts. Elliott was granted an injunction by Judge M.B. Gerry that blocked the sale of the mine.

In reply, both groups of lawyers questioned William Elliott's ownership of the mine. They charged that William Elliott had colluded with Ellen Jack to help her avoid payment of the debts owed to the attorneys. Due to the fraudulent nature of the sale, they requested that the injunction be lifted and Ellen's interest in the mine sold.

William Elliot defended himself from these charges by stating that he had a legitimate right to the mine. He stated that Ellen Jack owed only $350 to Gullett and Karr, not the outrageously higher figure that the lawyers claimed, and that Ellen owed nothing to Shear and Johnson. All parties would have to wait until August 1888 for further action to be taken on this case.

———

ELLEN'S TROUBLES WITH REDMOND Walsh continued when he reported her for pension fraud, a crime that could carry a prison sentence. The establishment of the widows' pension provided financial support, and it also meant that the government would investigate a person's private life if they were suspected of fraud. Each time Ellen had drawn her pension since 1876, she had signed an affidavit that she was not married. Redmond wrote a letter to the U.S. attorney for Colorado, Henry Hobson, stating that Ellen Jack had remarried twice. She married Jeff Mickey in 1881, and he had died by suicide. In 1882, she married R.D. Walsh, but she divorced him. Walsh called her "a hard case of a woman…doing lots of harm and encouraging chrime [*sic*] of all kinds." He also repeated his charge from the divorce case that Ellen was living with Frank Royer as husband and wife. He stated that she had persuaded several men to falsely swear on affidavits that she was not married. Henry Hobson forwarded the letter from Walsh to the commissioner of pensions in Washington, D.C., and asked for an investigation into the case.

Special Examiner J.T.H. Hall traveled to Gunnison in the dead of winter in January 1886. He took depositions from eleven local government officials and businessmen, who testified that they knew Ellen as being married to Jeff Mickey and then to Redmond Walsh. Hall tried to be as secretive as possible, but Ellen Jack was so "notorious" a person that the news of his investigation spread quickly in the small town of Gunnison. It didn't help Hall that the *Gunnison Review Press* reported that he was in town on a "secret mission."

Hall thought he had a clear case of perjury. Ellen Jack had signed affidavits saying she was not married. Hall had the testimony of eleven witnesses who stated that she had remarried twice. He found the records of her divorce case in Aspen, which he viewed as evidence that she had been married. Hall's recommendation to J.C. Black, the commissioner of pensions, was clear: her pension should be discontinued, and she should be "punished for her perjury."

The *Aspen Daily Times* reported that "the outcome of the present discovery cannot result in anything but conviction, and the pathway of the heroine of the Black Queen is evidently a thorny one from now on." The newspaper underestimated Ellen Jack's abilities to fight this charge.

Ellen Jack's trial took place in January 1887 in the court of District Judge Moses Hallett in Denver. Ellen hired a prominent lawyer in Denver, John D. Elliott (no relation to Ellen). Elliott had served as governor of Mississippi in 1853. He had moved to Colorado to practice law and operate mines in Irwin, north of Gunnison, during the rush in 1879–80. He had practiced law in Denver since the early 1880s.

U.S. district attorney Henry Hobson spent nearly $1,000 to prosecute a $66 fraud charge. He arranged for a clerk from the New York City Pension Office to travel by railroad to Denver to testify that Ellen Jack had received a pension during the time she was married. He subpoenaed Redmond D. Walsh from Memphis, Tennessee. The cost for travel and per diem expenses for the prosecution's witnesses totaled $801. Because Ellen Jack appeared to be poor, the government paid travel and per diem expenses of $125 for her witnesses.

Ellen Jack defended herself by stating that her marriage to Walsh was not legal because he was already married to another woman when she married him. Ellen had not known this information at that time. She first received this news in a letter from Annie Ganley, and after the two women met in Gunnison, Ellen was convinced that Ganley and Walsh were married. Ellen did not live with Walsh after discovering this information.

Annie's brother, Thomas Ganley, corroborated Ellen's story. Thomas said that Annie had married Redmond D. Walsh on August 10, 1881, and they lived together as husband and wife in Leadville. Thomas Ganley had traveled to Gunnison in 1882 and 1883, when he found out that Walsh had married Ellen Jack. Ganley believed that his sister was currently residing with Walsh in Memphis, Tennessee.

The twelve-man jury retired for deliberations on January 15, 1887, and on the same day, they returned a verdict of not guilty.

District Attorney Hobson explained the outcome to the commissioner of pensions: "There was no rational ground for her acquittal, but the jury refused to convict a woman in humble circumstances for what they considered to be a small offence against the government." As to why the jury acquitted her, a member of the jury told him that "she had crawled out of the smallest hole he had ever seen and one that no man would have gotten through," referring to the fact that Walsh had already been married to Annie Ganley when he married Ellen. Despite the acquittal, Hobson recommended that Ellen Jack be removed from the pension rolls.

Referring to the jury's verdict, Commissioner Black responded that "its acceptance by this office would be an outrage upon the government and a positive injury to the cause of justice." He felt that she deserved a "long term of imprisonment, as well as a heavy fine." Commissioner Black instructed Hobson to "proceed as expeditiously as possible. Have the criminal rearrested, and keep her closely confined in prison until she goes into the penitentiary."

Perhaps Ellen Jack's real crime was being an independent woman. She ran a boardinghouse, saloon and possibly a brothel; she owned a mine; and she had divorced Walsh. The original purpose of providing a widow's pension was to get men to enlist. But this program had fostered women's independence by providing a minimal level of financial support. Ellen's pension may have been one factor in her decision to divorce Walsh. Women who have some degree of financial independence are less likely to stay with an abusive husband. With the establishment of a widow's pension, the government now had an interest in policing personal relationships. A commissioner of pensions could terminate the pension of a widow if she engaged in "open and notorious adulterous cohabitation." The main issue in Ellen's case was whether she had been married to Walsh or not, but Walsh had also accused her of cohabiting with Frank Royer.

Although Ellen had won her court case, thus avoiding imprisonment, her ordeal was not over. District Attorney Hobson forwarded his case materials to the Pension Office in Washington. Her case was reviewed, and in July,

Ellen Jack was dropped from the pension rolls due to the evidence of her remarriage. Over the next couple of years, Ellen would seek to have her pension reinstated.

Ellen Jack's troubles with Redmond Walsh ended with this trial. By 1888, Walsh settled in East Chicago, where he engaged in the early development of the city by constructing streets and some of the first buildings. He became regarded as the "father of East Chicago." The *Encyclopedia of Genealogy and Biography of Lake County, Indiana* praised Walsh as a "man of distinct and forceful individuality, of broad mentality and mature judgment.…He has left an impress for good upon the industrial world. He earned for himself an enviable reputation as a careful man of business and in his dealings became known for his prompt and honorable methods, which win for him the deserved and unbounded confidence of his fellow men." This biographical sketch mentions Walsh's marriage to his first wife, who died in 1871. Their eight children had all died. There is no mention of Walsh's marriage to Ellen Jack or Annie Ganley. There is only an oblique mention of his mining activities in Colorado, which described that he "was interested in several diggings." Walsh died in 1918.

Ellen Jack and D.D. Fowler met as early as 1886, as indicated by Fowler having been listed as a surety on Ellen's bond in the pension case. An experienced mining man, D.D. Fowler started prospecting in 1863, and he had made discoveries in Idaho and Montana. He arrived in Aspen in May 1881 and discovered the Little Nellie Mine on East Aspen Peak. Each year, he returned home to his wife and children in Macon, Missouri, confident that one day, his prospects would make him a millionaire. In one of his letters from his prospecting trip in 1885, he warned of difficulties but remained optimistic:

To those who are comfortably situated at home, they are much better off than they would be here. A man lives a long time in this country and dies quick. True, there are many chances here for a man to make a big strike, but there are few who do. But all believe they will make the strike and are living in anticipation of that happy event.…My faith is yet positive of the bloom of the rose, that it will yet burst forth with its many flowers.…I am going to make that strike if it takes me five years to do it.

Fowler investigated the Black Queen Mine on behalf of investors. He traversed the high mountain passes, still covered in some snow, and arrived in Crystal, which was just emerging from its winter hibernation. Crystal's miners sought more investment dollars to develop the mines, and the town still needed a sawmill, a concentrator and more miners. Pleased with what he saw at the Black Queen Mine, Fowler agreed to buy the mine from Ellen Jack (for William Elliott), Young and Farnham for $25,000 in June 1887. The owners would retain title to the mine until Fowler delivered payment within the next year.

In the meantime, Fowler signed a one-year lease for the mine. After deducting the costs of transportation, workers and the smelter, Fowler would keep 87.5 percent of the profits, with the remaining 12.5 percent going to the owners. The mine owners signed the lease: Ellen Jack signed as the attorney in fact for William Elliott, George Farnham signed his name and Lark Young signed as the attorney in fact for C.W. Young.

The Black Queen revealed that its riches were worth fighting over. By the end of September, D.D. Fowler managed a force of six men, and in the next month, he broke into a rich pocket of silver ore. It surprised him that an experienced miner like George Farnham would have agreed to sell him "an undiscovered Comstock." Realizing his mistake too late, Farnham asked for "a future interest in the mine," but Fowler refused the request. It might seem that finding valuable silver ore would be a cause for celebration for all concerned, but instead, it instigated a new round of fighting.

Captain Jack suspected that Fowler was trying to cheat her. She made a surprise trip to the Black Queen. She arrived just as Benton's ore-laden burro train was starting down the trail on the steep mountainside. She drew her pistol and ordered the men to unload the cargo. They obeyed the mining queen.

Ellen saw Fowler coming up to the mine. She grabbed a rifle, stood in the entrance to the mine, and yelled, "Halt! This is my mine; no one takes ore; bond and lease forfeited; now git, or I'll give you the pure article from the gun!"

Ellen's retelling of this incident in her autobiography probably embellished the actual events. She said, "Benton tried to get his big gun out of his belt, then I sent a shot and took his ear off as clean as though it had been cut off and was going to send another, when he threw up his hands and yelled out: 'I am shot.' I sent a shot and shot two of the tips of Aller's fingers off; then he began to yell." These details of shooting off an ear and fingers are not confirmed by other sources.

Mrs. Captain Jack, starting for the mine. *Author's collection.*

Ellen Jack gained nationwide notoriety when the story appeared in several newspapers around the country, including the *Rocky Mountain News* (Denver), the *Burlington Daily Gazette* (Iowa) and the *Sun* (New York City). The *St. Louis Globe-Democrat* titled its article: "She Pulled a Gun and Held the Ore."

This incident would also be recalled years later, when newspapers printed columns about events that had occurred in previous decades. The *Gunnison-News Champion* remembered the excitement two decades later in 1907: "The famous litigation over the Black Queen Mine started with a gun play in which Mrs. Ellen Jack invited D.D. Fowler of Missouri, the lessee who had just made a rich strike, to go down the trail." In 1914, the *Elk Mountain Pilot* at Crested Butte reprinted a news item in which it recounted, "Mrs. Jack came out of the cabin with a gun, telling them she owned the mine and would not allow them to move the ore....As we go to press, Mrs. Jack still holds the fort." A week later, the *Elk Mountain Pilot* told the resolution of the story: "Benton's burro train of fifty-six jacks was loaded with Black Queen ore, the difficulties at this mine having been overcome, and started for the railroad at Crested Butte."

Mrs. Captain Jack brandishing her gun. *Author's collection.*

D.D. Fowler described the incident quite differently from his perspective, saying that "they came in mob force, headed by a redheaded woman, one of the owners. The crowd had a proclivity for jumping mines and shooting. I thought I would have to surrender the mine, but believing that many a battle had been lost on a bluff, concluded to stay and take chances on law." Fowler called on his business associate Sheriff Doc Shores, whose help he had enlisted the previous spring. Fowler paid Doc Shores for his services rendered in the Ellen Jack case.

Fowler had "the Amazon," which was how the *Gunnison Review Press* described Captain Jack, "arrested for threatening his life." Doc Shores arrested Ellen, and he took her by wagon to Crested Butte, where they stayed overnight, and the next day, they took the train into Gunnison.

Ellen was charged with assault with intent to kill. Justice Piper ruled in Ellen's favor. The justice explained that Ellen had proven her claim that Fowler had violated the terms of the lease, and regarding the assault charge, he declared that "Mrs. Jack had a right to protect her property." It may have helped Ellen's case that James Piper's family socialized with

Lark's and John Young's families and had a grocery business together for a time.

D.D. Fowler called Piper an "ignorant justice" in a letter printed in his Missouri hometown newspaper.

Instead of settling differences with her gun, Ellen faced D.D. Fowler in court. Fowler filed a suit against Captain Jack and her co-owners for $5,000 in damages, claiming that they had prevented him from working the mine. Judge M.B. Gerry allowed Fowler to continue mining the ore, but neither party could ship or sell it until the case was settled in the regular term of the court. Confident that he would win the suit, Fowler continued mining throughout the winter.

The two parties worked out their differences. In June 1888, Captain Jack and her co-owners entered into an agreement with D.D. Fowler and John T. Johnston, an investor from Chillicothe, Missouri, near Fowler's home of Macon. Johnston had made money in manufacturing tobacco products, and he had served as the mayor of Chillicothe. The agreement stipulated that the owners would sell the mine to Fowler and Johnston on November 1, 1889, for $25,000. Until then, Fowler and Johnston would sell the ore; deduct expenses of mining, shipping and smelting; and give the remaining profits to the owners. Fowler obtained possession of the ore that had been stored at a warehouse in Gunnison. He could now sell it, provided that he gave the owners their share of the profit. The "Black Queen difficulty" appeared to be over, and the company was shipping ore to the smelter in Pueblo.

Captain Jack and her co-owners continued negotiations with interested investors. In September, John W. "Genial Jack" Bowlby, a wealthy investor from Kansas City, Missouri, traveled to Crystal to see the mine. Before Bowlby left Kansas City, he made out a will, stating, "I start for Crystal, Colo., today. In case I should die, or should be killed, or, in other words, switched for the other shore—heaven, I hope—I want all my [bank] deposits…to be paid over to my wife, Lucy Bowlby. This I do for her protection, as we have no children."

In Crystal, the portly forty-seven-year-old Bowlby began having heart trouble, possibly aggravated by the high elevation. Compared to the eight-hundred-foot elevation of his hometown, spending a week at nine thousand to ten thousand feet could spell trouble for someone in poor health. Al Johnson's brother, Fred, accompanied him on the return trip to Gunnison as they went up over Scofield Pass and then rode the train from Crested Butte. Upon their arrival in Gunnison, Dr. Rockefeller was called to administer aid. Mr. Bowlby spent the night at the best hotel in town, the La Veta. The next morning, Bowlby insisted on returning home, casting aside the protests of his companions, but as Fred Johnson assisted him onto the Pullman car, he collapsed into Johnson's arms, dead from a heart attack.

In explaining Bowlby's death, the *Gunnison Review Press* described him as "a very fleshy man and full blooded—apparently of apoplectic tendencies. He weighed 250 pounds and yet was not above the average height." Fred Johnson took Bowlby's body back to Kansas City for burial.

Ellen Jack won back her widow's pension. In September 1888, Ellen won a ruling in the Gunnison County Court that declared her marriage to Walsh null and void. Her lawyer, Henry Karr, wrote to the commissioner of pensions for reinstatement. Karr stated that he believed Ellen had not been married to Jeff Mickey and that those who had testified previously had been merely stating their suspicions.

Ellen begrudgingly paid her debt to Gullett and Karr from previous cases, which had swollen to $1,754 due to a high rate of interest. Referring to the debt, Ellen said, "I should call it stealing. He took it [the money] under the pretense of attorney's fees." She also criticized the appearance of propriety among such people who practiced usury while claiming to be followers of Jesus:

> *Go into the church and look at the congregation, all well-dressed people and gray-haired old women and bald-headed old men. They pray to Almighty God to send them prosperity. He has already sent them poor brothers who have to have money at 2 per cent per month, or as much more as they can hold them up for, and practice usury from Monday morning to Saturday night; then Sunday morning, their piety starts and ends Sunday night, and they have done their duty.*

Ellen Jack accused Henry Karr of stealing, but the community held him in high regard. In January 1889, Henry Karr died from pneumonia at age forty-eight. His law partner, John Kinkaid, made the funeral arrangements. The mourners in the overflowing church heard Reverend Fueller memorialize a leading member of the bar in Western Colorado, saying, "We respect talent, we admire public spirit, we value integrity and usefulness, but when to these qualities is added a generous, hearty, affable, jovial, cheerful manner and disposition, then it is that our affections are strongly enlisted. It was the possession of all these qualities that so endeared Judge Karr to this community."

Captain Jack and her co-owners had been battling Al Johnson over ownership of the Black Queen Mine for over two years. In September

1887, Johnson had sold the Excelsior claim to the lawyer C.J.S. Hoover for one dollar. Hoover and U.S. deputy mineral surveyor Robert Sterling had claims together on Crystal Mountain, and they shared office space in Crystal. Hoover prepared patent papers for mine claim owners. A few months later, Hoover sold a one-third interest in the Excelsior to Alfred Oskamp, an investor from Cincinnati. Oskamp became a part owner at the recommendation of his friend from Cincinnati, Robert Sterling.

The trial had taken place in August 1888 in the District Court of Gunnison County. Several people from Crystal testified for Hoover's side, including Al Johnson and Robert Sterling. The twelve-man jury included Ellen's old friend John Roberts and Lark Young's friend Joseph Gavette. The jury deliberated for less than an hour and returned a verdict for Ellen Jack (for William Elliott), Young and Farnham. They would retain ownership of the Black Queen Mine. Lark Young, George Farnham and several of their friends celebrated their court win with an oyster supper at Gavette's restaurant.

In April 1889, C.J.S. Hoover appealed the case, but Ellen Jack (for William Elliott), Lark Young and George Farnham won this round as well. Hoover filed an appeal of the case in the Colorado Supreme Court. He believed that the juries in both trials had not taken the time to consider the evidence or read the judges' instructions, having returned with their verdicts in under an hour. Hoover believed that he could not get a fair trial in Gunnison because the residents of Gunnison County were prejudiced against him, and he charged that Ellen Jack, Young and Farnham had an undue influence over people in the county. John Johnston told Hoover that "it did not matter if he took his case to the supreme court and there had the judgment reversed," because Johnston and his associates "could always win the case before a jury in Gunnison County."

During this trial, questions arose concerning the identity of William Elliott. Farnham testified that he had not met William Elliott but had heard that he was one of the owners. Ellen testified that William Elliott was her brother and that she had sold her interest in the Black Queen to him. Rumors had swirled around Gunnison for some time about the identity of "William Elliott." Up to this time, Elliott had not appeared in person to present his testimony. His lawyers had filed the court documents for him.

A turning point in this case occurred in April 1889. Because William Elliott was not a resident of Colorado, the court issued a cost bond, which guarantees payment of court costs. If Elliott failed to put up the cost bond, a judgment would be rendered in favor of Shear, Johnson and Shores. Elliott's lawyer, John Kinkaid, filed the cost bond on May 3, one day late.

John Kinkaid told Sprigg Shackleford that William Elliott would appear in court to testify. Shackleford responded that "credulously and doing great violence to his better judgment, [he] fondly hoped and trusted the word of Kinkaid that the mythical Elliott might appear."

At the end of June, Sprigg Shackleford and his cocounsel Sam Crump made the accusation that there was no such person as William Elliott. They charged that Frank Royer had pretended to be William Elliott in obtaining power of attorney for Ellen Jack. Shackleford had just received information that a relative of Ellen's deceased husband, Mrs. M.R. Quackenbos from New York, was willing to testify that Ellen did not have a brother named William Elliott. Ellen had left her daughter, Adeline, with Margaret Quackenbos when she had left for Colorado. Another witness, Sam Brust, was willing to testify that Frank Royer was using the name William Elliott. Brust had owned a cigar store in Crested Butte in the early to mid-1880s, and he now lived in California.

Shackleford had been unable to find anyone who had seen William Elliott. He complained that he had not "been regaled by a sight of the promised appearance of the supposed plaintiff [Elliott] in this action and has had no opportunity to take the testimony of the mythical Elliott." Shackleford and Crump believed that Ellen had constructed this fictitious person for the purpose of defrauding Shear and Johnson. Because Elliott did not exist, Ellen Jack had not sold her mine to him. The lawyers pressed this issue further by requesting that the court order Ellen and her lawyers to provide Elliott's address and information about his identity. They requested that Elliott be required to appear in court in person.

The court ruled in favor of Shear, Johnson and Shores in August, due to William Elliott's failure to put up the cost bond. Judge John C. Bell declared as fraudulent Ellen Jack's supposed sale to William Elliott. The ruling lifted the injunction against the sale of the mine. The mine could now be sold to satisfy the debt. The outcome of this case put pressure on Captain Jack to sell her mine before Sheriff Doc Shores could auction it off. At a sheriff's sale, the mine might be sold for the mere cost of her debts, which might leave Captain Jack with little to no proceeds at all for her mine.

D.D. Fowler and John Johnston still held the lease on the Black Queen, and they increased production in the summer of 1889. Manager Pat Moore employed fifteen to twenty men, with a payroll of $2,500 per month. The main incline shaft had been sunk to a depth of 250 feet. At this depth, the water came in fast, so John Johnston and George Farnham put in a steam hoisting works to dewater the mine. Forty burros, each one loaded with two hundred pounds of ore, trod along a thirty-five-mile route that followed the Crystal River downhill to Carbondale, and from there, the railroad transported the ore to Denver.

It is usually difficult to know if a mine was profitable or not. Mine owners might have downplayed the value of the mine if another party was trying to gain control of it, as was the case with the Black Queen. According to court testimony, even with all the activity in 1888–89, profits continued to elude the mine owners. The lessees, Fowler and Johnston, had spent $2,000 on machinery. Their expenses totaled $25,450, and income from the ore brought in $10,950, which left a net loss of $14,500.

As the saying goes, "It takes a mine to make a mine." Mining has long been regarded as akin to gambling. The hope of hitting it big kept people involved in mining, despite of the small odds of finding a bonanza. More money may have been invested than the owners gained in return, with estimates ranging from five dollars to one hundred dollars invested for every one dollar gained.

On the other hand, mine owners would typically play up the value of their mines when they were trying to sell or attract investors. An article in the newspaper of John Johnston's hometown of Chillicothe, Missouri, reported that the bond of $25,000 and the cost of machinery was paid off in less than a year, which seemed to contradict the previous court testimony. The article also stated that there was $60,000 worth of ore in sight.

Negotiations for a sale had been taking place for some time. In the *Crystal River Current*, Al Johnson did not mention Captain Jack by name but stated, "Col. [John] Johnston, with a meekness that would severely test the patience of a saint, was negotiating with the owners." John T. Johnston, Archie McVey, W.H. Mansur, Eugene P. Shove, C.W. Shores and John Kinkaid incorporated the Crystal River Mining Company in August. The company had offices in St. Louis and Chillicothe, Missouri, and in Crystal, Colorado.

C.W. Shores personally benefitted from his actions as sheriff by being included as one of the incorporators of the Crystal River Mining Company. He had applied pressure on Ellen several times by trying to sell the Black Queen to satisfy judgments against her. He had been paid by the Crystal River Mining Company for services rendered, such as bringing Captain Jack into Gunnison after she had threatened D.D. Fowler. Shores owned the adjoining Fargo Lode.

John Kinkaid, another one of the incorporators of the Crystal River Mining Company, worked as an attorney for the company, and he helped with negotiations for the sale. Kinkaid had also been Ellen's lawyer in the case against Shear and Johnson. In that case, he had failed to complete the paperwork on time, which led to the ruling against Ellen that put pressure on her to sell the mine. It seems questionable whether Kinkaid could fairly represent Ellen's interest when he also worked for the Crystal River Mining Company.

On November 2, Ellen Jack and her co-owners sold the Black Queen Mine to the Crystal River Mining Company. Ellen Jack received $5,000 for William Elliott's half interest, George Farnham received $4,500 for his quarter interest and C.W. Young received $4,600 for his quarter interest, according to Gunnison County documents. Ellen said that she enlisted the aid of John Kinkaid to keep additional proceeds from the mine for a time while she dodged the debts claimed by Shear and Johnson. She would have a hard time collecting any additional money from the sale.

In her autobiography, Ellen said that she sold the mine for $25,000, with $5,000 down and the balance to follow. Ellen did not mention her co-owners or Frank Royer, nor did she mention anything about the involvement of William Elliott in the Black Queen. She did say that William was her brother's name and her father's name, which is confirmed by census records.

Ellen Jack faced yet another challenge to keeping the proceeds from the Black Queen sale. Ellen charged that certain parties—she did not know who—tried to defraud her of the money owed to her. Perhaps the lawyers whom Ellen had fended off with her creation of William Elliott would now try to use that fictitious character for their own benefit.

Several reports surfaced that said there was an actual person named William Elliott. On December 6, 1889, lawyer Sam Crump said that he saw "William Elliott" with Ellen Jack in Gunnison. Elliott was in town conducting business with the Thomas Bros. and Wegener Law Firm. Thornton Thomas confirmed this information, saying that he had met with William Elliott in person. James Piper also stated that Ellen Jack had come into his store, and she had introduced William as her brother. Perhaps the mythical William Elliott existed, or maybe someone was just pretending to play the part.

An unexpected claim surfaced from a man who claimed to be William Elliott. Thornton Thomas received a letter in February 1890 from attorney Samuel Kingan of Ogden, Utah, claiming that he represented the interests of William Elliott. Samuel Kingan claimed that Ellen Jack had tried to enlist the help of this William Elliott to avoid paying Shear and Johnson and that she owed him $1,500 for his assistance. If Ellen would send the money, Elliott would refrain from exposing her fraudulent scheme.

Sam Crump confirmed that he had seen the same "William Elliott" in Ogden, Utah, at the law office of Samuel Kingan. At that office, "William

Elliott" told Crump that his real name was Curtis and that Ellen Jack had enlisted his help to avoid paying Shear and Johnson.

In March, Kingan requested that the Colorado Supreme Court list his name as the attorney for William Elliott. Elliott granted power of attorney to Kingan, revoked the power of attorney he had given to Ellen Jack and severed ties to his previous attorneys. He further stated that Kingan was authorized to collect money from the Crystal River Mining Company on his behalf. It appeared that this "William Elliott" was trying to collect the further payments for the Black Queen Mine that was owed to Ellen.

Thornton Thomas defended Ellen. He stated that the person represented by Kingan was an imposter, and he accused Kingan, Shear, Johnson and Shores of knowing that this person was an imposter. They may have promised this person something in the event of a favorable outcome of the case. Nothing further was ever heard from this supposed "William Elliott" or Samuel Kingan.

Thornton Thomas successfully ended these fraudulent schemes, but he extracted his own pound of flesh. He obtained payment for his legal services by acquiring Ellen's interest in the Lead Chief and Bonanza King Mines. In 1886, William Elliott paid $2,000 for a part interest in these two mines, which were located across the Crystal River Valley from the Black Queen Mine, high on Whitehouse Mountain. William Elliott sold the properties to Ellen's daughter, Adeline Snevily, in November 1890. But using other people's names in place of her own didn't work this time for Ellen. Thornton Thomas obtained a judgment for $223 against all three people—Ellen Jack, Adeline Snevily and William Elliott. Thornton Thomas bought the properties at a sheriff's sale in June 1892.

Ellen said that she paid her debts but that while she was waiting for the rest of the money from the Black Queen sale, the lawyers were scheming about how to get a hold of it. Their trumped-up charges from Ellen's previous cases added up to $5,000. She directed John Kinkaid to hold money from the sale so she could avoid paying Shear and Johnson and the other lawyers. But with John Kinkaid playing both sides of the fence in the sale of the Black Queen, Ellen should not have trusted him with this scheme either. She would continue to face battles with Kinkaid over obtaining the rest of her money.

A description of Captain Jack's predicament appeared in the *Queen Bee*, a Denver newspaper that was "devoted to the interests of humanity, women's political equality and individuality." Under the headline "Prosecution of Businesswomen," an article gave a strongly worded, accusatory account against unnamed men concerning Captain Jack's struggles over her mining properties:

There is a band of masculine thieves living at Gunnison City, Colorado, who have clubbed together to rob Mrs. Captain Jack, a bright little woman who makes her money legitimately, pays her bills and attends to her own business. Mrs. Jack is the only female mining expert in the United States. This woman has the ability to make money and compete with the best masculine brain in the country. Mrs. Jack is an educated woman and an excellent shot, she takes her horse, wagon and camping outfit goes alone prospecting, no one molests her until she proves her works by a sale of valuable mining property. Then these masculine hounds will lay for her with every process of man-made and man executed law and rob her of thousands. It is a common saying in Gunnison that this band of official thieves would have gone hungry but for the blood money paid them by Mrs. Jack. Those men have the same country needing developments in which this little mite of a woman makes her thousands; they loaf in their offices, drink at the bar, tell stories in their leisure moments, while this woman gives her time to the study of her subject, sober application, hence enviable success. Now, it would be better for these men to act as cooks, there are plenty of hotels and restaurants needing good and faithful cooks; this would bring them more glory than the robbery of women. Mrs. Jack would be perfectly justified making a target of this band of thieves. Lately, in a suit brought against her by a person who had forged her name that he might claim half of a piece of mining property which she owned, every point of law was in favor of the woman, yet the decision of the court was against her because this was the only way to get the costs paid. The man forger not being able to pay the costs. Colorado is getting a fine reputation for its treatment of women. Mrs. Jack's is not the only instance of brutal injustice and cruelty to women.

ELLEN JACK AND FRANK Royer do not appear to have had any other contact with each other after the sale of the Black Queen Mine. Frank Royer achieved much professional success as a mining engineer. In 1899, he graduated from the School of Mines in Golden, Colorado. The *Colorado Transcript* described Royer, the son of a carpenter, as "a striking example of what a poor boy can accomplish in the West." He earned the respect of mining men and had many top offers of employment. He became the manager of the famous Dolores Mine in Mexico for the noted mining magnate John B. Farish.

Built in 1893, the Crystal Mill harnessed hydropower to produce electricity used in mining operations. *Jane Bardal, photographer.*

Some of the men who led the Crystal River Mining Company would end up with considerable wealth from the enterprise. President John T. Johnston's estate included $54,000 worth of stock in the company. D.D. Fowler's estate was reportedly valued at $200,000 at the time of his death in 1899. Toward the end of Fowler's life, several people accused him of insanity as they vied for control of his wealth.

Al Johnson did not benefit from the Black Queen's riches. He died at age thirty-seven and was buried in his hometown of Montreal, Quebec. His wife, Kate, and his brother, Fred, continued living in Crystal. At the time of his death, Al Johnson was in debt to the bankers R.G. Carlisle and John Tetard. He had secured the promissory note with his store, printing office and part interests in seven mining claims. The bankers showed no mercy to his widow, Kate Johnson, and with Sam Crump as their attorney, Carlisle and Tetard won a judgment against Albert Johnson's estate, and they also acquired all his property.

The Sheep Mountain Tunnel and Mining Company hoped to unlock the riches of the Black Queen and many nearby mines by driving a tunnel to intersect the ore bodies. With its construction started in 1891, after four years, the tunnel had been driven 3,600 feet into the mountain, with no significant amount of ore found. The failure of the Sheep Mountain Tunnel and Mining Company resulted in a sheriff's sale in 1902. The company's power plant would serve as a lasting monument to the search for silver.

5

CAPTAIN JACK GOES PROSPECTING IN OURAY AND WHITE HILLS

*E*llen said that she made a lot of money from the sale of the Black Queen Mine, but the lawyers had to get paid first before she could obtain a clear title to sell the mine. She railed against the lawyers who gouged her with their fees. She said, "I was so disgusted that I left the place and went to Ouray."

Named for the Ute chief, the town of Ouray was located about one hundred miles southwest of Gunnison along the route of the Denver and Rio Grande Railroad, which had just reached the town the year before. By the time of Ellen's arrival in late 1889, Ouray was a well-established mining town of two thousand people. Several stately brick buildings adorned the town, including the Wright Opera House, the Beaumont Hotel, a schoolhouse and the Miner's Hospital. Four churches, two newspapers and twenty-six saloons served the town's residents.

Ellen bought a couple of mining claims near Ouray. Acting as attorney in-fact for her twenty-one-year-old daughter, Adeline Jane Snevily, Ellen bought a half interest in the Alta Rico claim in December 1889. Adeline lived with her husband in Brooklyn, New York. At age eighteen, she had married the thirty-eight-year-old Charles Edward Snevily in 1887. Charles Snevily was the son of John Snevily and Harriet Jack, the late Charles Jack's sister, which meant that Charles Snevily and Adeline were first cousins. Charles was a wealthy broker, and Adeline was a well-known society woman. The couple lived on Flatbush Avenue, across from the Midwood Club, where Charles was a leading member and Adeline held positions on the best committees.

Ouray. *Author's collection.*

The Alta Rico claim was located about a mile north of town on Gold Hill in the Uncompahgre Mining District. Numerous prospects on Gold Hill, also called the Gold Belt, had produced thousands of dollars' worth of gold in 1889. The American and Nettie Mine had sent down carloads of ore, and at least ten nearby claims caused much excitement as well. In the same week that Ellen bought her claim on Gold Hill, the American and Nettie Mine sent down between 150 and 200 sacks of gold ore every day. It looked like Ouray would soon become known for its gold mines, in addition to the already prominent silver mines.

Ellen sold to Adeline the Thunderbolt claim, located west of town up the steep Oak Creek drainage. A few years earlier, several gold prospects on Oak Creek had shown some initial promise, and two prospectors had sold their claims for $18,000 to parties from St. Louis. Because the ore came from the same formation as the Gold Belt, prospectors and investors hoped it would bring similar yields.

In July 1890, Ellen bought a beautiful Victorian-style two-story house that was located one block east of Main Street. Ellen also bought the contents of the house from the previous owner, which included a piano, table, beds, chairs, lamps, window shades and kitchen utensils. A bay window on the south side and a porch on the west entrance ornamented the house. The surrounding yard contained flower and strawberry beds, currant bushes and fruit trees.

Ellen said that she "found a good showing of silver up Bear Creek Falls." In September 1890, she obtained a half interest in a lease on the Little Maudie Mine from Evans Willerup. The mine had been discovered in 1881 by Henry Sefton and the Old Reliable Mining and Reducing Company. Sefton and his company also patented other claims nearby: the Mountain

Queen, the Big Fraud and the Little Fraud Lodes. An adjacent claim, the Sivyer, owned by Judge A.L. Sivyer, had produced $16,000 worth of silver.

To get to the mine from Ouray, Ellen would have spiraled up the serpentine switchbacks on the Ouray and Silverton wagon road heading south, high above the Uncompahgre River drainage. The wagon road intersected the steep and narrow Bear Creek drainage. At that point, Bear Creek rushed through a constricted passage, plunged 227 feet onto the rocks below and joined the Uncompahgre River. Roadbuilder Otto Mears built a toll bridge over the top of Bear Creek Falls. On the south side of the Bear Creek drainage, a pack trail scampered uphill for about one mile to the Little Maudie Mine, perched 400 feet above Bear Creek.

By November, Ellen had two men working at the mine, and she had shipped four carloads of high-grade ore. The *Solid Muldoon* said that Mrs. Jack was "certainly a woman who deserves great credit for her mining knowledge and pluck." David Day had a reputation for being a controversial, scrappy newspaperman who loved a good fight. Perhaps he saw Ellen as a kindred spirit. In the *Solid Muldoon*, he sang Ellen's praises as a female miner:

> *Chicago's Evening* Post *credits Mrs. Emily Knight of Tacoma and Mrs. Hensley of Castle, Montana, with being the only "female miners in*

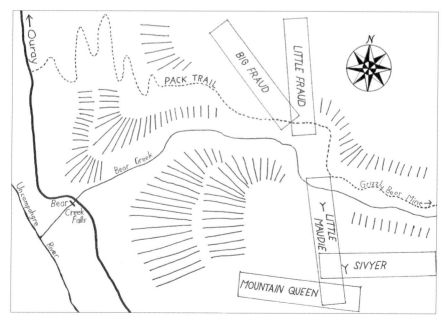

Bear Creek Mines, showing location of the Little Maudie claim. The pack trail on the north side of Bear Creek was built in 1896. *Drawn by the author.*

Bear Creek Falls. Note the wooden bridge over the top of the falls. *Author's collection*.

the world." We beg to differ, as Ouray has in the person of Mrs. Capt. Jack, the best, shrewdest and most successful all around female miner and prospector in the business. She has located, worked and sold a number of properties and is now east engaged in placing very valuable claims. Mrs. Jack takes life as it comes and when occasion requires can paralyze the opposition with silk, satin and diamonds, or make a burro feel at home in her overalls and slicker. She attends strictly to her own business, gets the best of the boys when she can in a business way, and makes and spends lots of money—as all miners do.

Ellen experienced financial trouble once again. Leo Erdman, the previous owner of Ellen's house, won a suit against her over a disagreement about the price of the piano. Ellen borrowed $470 from John Kinkaid, who had relocated his law practice to Ouray. He sold the promissory notes to the First National Bank of Ouray, and the bank won a suit against Ellen to recover the money. Also, the San Juan Hardware Company won a suit against her for $305. The sheriff scheduled a sale of her house for January 30, 1892.

Ellen said that the mine near Bear Creek Falls paid her expenses for a while. She had a buyer who offered her $60,000 for the mine, but before the deal could go through, the price of silver plummeted. The buyer backed out of the deal, and she lost money on the mine.

Owning mining claims close to valuable mines does not guarantee success. Ellen owned claims in the Gold Belt and along Oak Creek, but neither of these gold mining claims produced the riches she sought. With no fortune found in Ouray and the bill collectors closing in, Ellen returned to Gunnison.

Ellen tore down her three Jacks Cabins, which were considered shanties by this time, and she used the lumber to build two dwellings in April 1892. She spruced up her property by putting up a fence and planting two evergreens. Ten years earlier, Captain Jack and others had prospected in the Kezar Basin southwest of Gunnison, but their efforts soon fizzled out. Now, prospectors renewed their interest in this area, and over the summer, a few of them made promising discoveries.

Captain Jack's old friend Susan Bryan sold a mining claim for $750. Susan Bryan went to Durango in December to find her fortune, with the *Gunnison News* stating that "she numbers everyone in town as her friends" and that all wished her well. She returned to Gunnison at some point, for in 1893, she staked a claim south of town with her daughter, Mary Anderson. The pair of women staked two additional claims in the Goose Creek Mining District, with Susan's daughter-in-law, Hester Bryan, staking a claim as well.

The Little Maudie was on the slope on the left side of the photograph. Note the pack trail cut into the mountain (*center of the photograph*). *Jane Bardal, photographer.*

But neither Ellen nor any prospectors she had working for her found anything in the Gunnison Gold Belt in the summer of 1892. Ellen said that in Gunnison, "there was nothing for me to do but watch my property, and that was too tame for me, so I decided to go west."

Captain Jack ventured to the White Hills of northwestern Arizona in search of a silver bonanza. The rush had begun in 1892, when a Native named Hulapai Jeff showed prospector Henry Schaeffer the location of a silver deposit. By late June, two hundred men and a few women braved the sizzling sun in the sandy Arizona desert to stake claims and set up a town. Merchant W.H. Taggert couldn't construct his store quickly enough to avoid some of his canned goods bursting open, his eggs turning rotten and his butter melting due to the scorching heat. Four saloons provided "red ruin" to the thirsty prospectors, three restaurants provided quality meals and boardinghouses and tents provided shelter. Schaefer sold his claims for $250,000 to Denver mine investors in October. Money could be made in the new bonanza silver district.

Captain Jack set off on her journey with ranchman Liborius Ahrens, whom she referred to as "Luey," and a neighbor widow woman with three boys. The party traveled by wagon through St. George, Utah, after which they slogged through the desert sand and crossed what Ellen called the "Verden River" (probably Virgin River). She mentioned traveling through "Bunkesville" (probably Bunkerville, Nevada, just over the Utah border), and then the party forded the Colorado River by ferry boat into Arizona Territory. She sent Luey and the woman and her boys ahead to White Hills, Arizona. Ten days later, Ellen rode a horse by herself over the cactus-filled, sandy desert to White Hills, which referred to both the town itself and to the series of ashen white, low-lying hills that surrounded the town.

By the time of Ellen's arrival, the town's population had risen to four hundred, and fifteen mines dotted the hills within a mile of the camp. The leading mine, the Grand Army of the Republic, turned out ore with one thousand ounces of silver per ton, and it had $100,000 worth of ore in sight. Denver capitalists David Moffat and R.T. Root had started negotiations to purchase several mining claims.

Captain Jack attended to several sick men at the hospital. She said that four men died from arsenic poisoning, but that she saved some of the men with her treatment. The *Mojave County Miner* reported that Dan'l Steel, age forty-eight, died from arsenic poisoning in mid-March in White Hills. A physician and two nurses attended to the dying man, but Ellen Jack is not mentioned by name in the newspaper. The doctor attributed the poisoning to the air in the mine, but other miners blamed the drinking water.

Ellen did not like desert living. She slept on the ground one night in camp and awoke to a rattlesnake underneath her. Fortunately, it did not bite her. Another night, she awoke to lizards covering her tent. The only available water was warm and nasty and made her long for the clear water of Colorado. Ellen explained the cause of the dry desert: "There is another triangle, and its law is repulsion, retention, suppression, and attraction, for what holds large quantities of water and snow which falls to the earth. There is nothing neutral, and it attracts from earth and is retained in the clouds until the right attraction draws it down."

Ellen took the stage to Kingman, Arizona, a trip that took ten hours to cover the sixty-mile desert route. The *Mojave County Miner* of April 1, 1893, listed Ellen E. Jacke as a hotel guest at the Hubbs House in Kingman, Arizona. The newspaper also briefly stated that a "Mrs. John Jack [probably Ellen Jack] of Denver, Colorado, who has been out to the White Hill country for some time past, passed through Kingman on her way to Denver. Mrs. Jack is interested in mining and will return to this place in a short time." But Ellen did not return to White Hills; instead, she returned to the cool, clear water of Colorado, where she would seek her fortune in the Gunnison Gold Belt.

Ellen ended her autobiography with this adventure in White Hills. She closed her book with what she called a "prophecy of the Fated Fairy," in which she stated her opinion on religious issues that she pondered as she sat on a fallen Joshua tree in the desert. In the moonlight, the Joshua trees looked like men and animals, with their yucca leaves shooting out in all directions from the ends of spindly branches. She reflected on a passage from the Bible: "If your wife offend thee, scourge her with a whip." Ellen stated her view:

> But women's intelligence soon came to their aid, and they would not let their children go to church or school to be taught such things, and when the priests tried to frighten them to send their children and go themselves, they stood firm and said: "No, we will call on God to save us from the sting of the lash when our husband is inflicting it on us, and our conscience tells us that God never made man so big and strong to use that strength on the weaker women with a whip, and we shall not have our children taught to do such things or our girls to submit to being whipped by their husbands."

She comments on several other topics, and she ends her prophecy by saying, "So, cheer up, for the aura light is breaking through the dark circle of apprehension."

6

PROSPECTING IN THE GUNNISON GOLD BELT

$\mathcal{E}$llen Jack was still having trouble collecting the money from the sale of the Black Queen Mine. In April 1894, she filed a lawsuit against John Kinkaid, whom she had employed to collect her money. Kinkaid had obtained $14,000, but he had paid Ellen only about half of that amount. Kinkaid had attained prominence as a lawyer, and he reportedly made $25,000 a year. In the early 1890s, Kinkaid and his wife moved to Denver, where Kinkaid organized a law firm. Mary Holland Kinkaid would become the deputy to the state superintendent of public education in 1896. Ellen would have her work cut out for her in pursuing a lawsuit against such a well-connected attorney. This case would be contested for several more years.

Ellen joined the rush of prospectors to the Gunnison Gold Belt in the spring of 1894. Ellen and many other prospectors had discovered gold in this area south of Gunnison in the early 1880s, but interest soon faded when not much of the yellow metal was found. Maybe what the area needed was further development to uncover the great mother lode. The silver crash in 1893 had lowered the price of silver, and many mines had closed, throwing thousands of miners out of work throughout the state. Coloradoans needed a shot in the arm with new gold discoveries. The Cripple Creek District had given the state a boost, with its gold production topping $2 million in 1893.

Perhaps the Gunnison Gold Belt could also be another overlooked locality that held great gold deposits. The grassy, rolling hills didn't look like the craggy, mountainous topography of most of Colorado's other mining districts, but the terrain did resemble the Cripple Creek region. The Gold

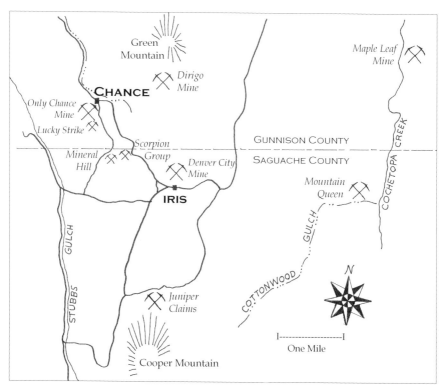

The southeastern part of the Gunnison Gold Belt, showing the locations of major mines and Ellen Jack's claims. *Drawn by the author.*

Belt encompassed a two-hundred-square-mile area that stretched for thirty miles from west to east, with Dubois on the west end and the Cochetopa area on the east end. Percy Williams, the editor of the *State School of Mines Scientific Quarterly*, reported that the ore samples from Dubois were even better than the samples he had taken at Cripple Creek.

Captain Jack staked two adjoining claims on June 13, the Scorpion and the Sunday Morning. By the end of July, she had located two more nearby claims, the Scorpion No. 2 and the Scorpio. The claim names reflected Ellen's astrological sign of Scorpio. She had an assay performed on the ore from her Green Mountain Mining District claims that returned a value of $600 in gold per ton. Now that she had a good showing from her prospects, she had to defend herself from claim jumpers. A mining partner of Charles Grant, Charles Bennett, along with other men, located an outcropping only 320 feet from Ellen's discovery shaft.

She told him, "Get off of my ground."

Mrs. Captain Jack packing to prospect. *Author's collection.*

He replied, "We're here to stay."

Ellen went to her camp and returned with several guns, telling the men, "My .44 has just made me a match for your gang. If you don't leave, I'll arrange it so that you have to be carried!"

The men left quickly. On Monday, August 13, Saguache county sheriff Baker went to the mining camp of Union Hill and arrested Ellen Jack for drawing a gun on Charles Bennett. Sheriff Baker took Ellen to Saguache, where she was arraigned before a justice of the peace. Her trial was scheduled for January.

An investor who had been in Union Hill at the time of the incident repeated the story to the *Rocky Mountain News*, which printed the byline: "A Woman's Nerve." The article described that "a character in the camp is Captain Jack, a woman prospector, who holds down four claims with a Winchester." This story made it into several newspapers around the country, including the *Black Hills Daily Times* in Deadwood, South Dakota, which said that "the woman is notorious throughout the country."

Ellen Jack was back on her claim again before the end of August. The *Queen Bee* provided further commentary on the incident:

> *Captain Jack is a terror to jumpers, and in order to save bloodshed, they will keep her shut up. If there is another striking war where it is necessary to proclaim martial law, it would be a good plan to enlist Captain Jack. Let her have command of a company at least. Lots of people would put up money upon Captain Jack's administration of affairs. It will be interesting to watch how the little Captain pans out in Colorado politics, for her sex have been emancipated but about one short year. Let it be remembered, if Captain Jack perishes, she will sell her life as dearly as possible. Men are simply absurd, or they would let her alone and fight professional pugilists and small dogs.*

The statement about emancipation referred to the fact that Colorado men had voted in favor of women's suffrage in 1893, and three women had been elected as state representatives in 1894.

The day after Christmas, Ellen traveled to Saguache to stand trial for assault on Charles Bennett. A mild start to the winter, with only a foot or two of snow along the route, benefitted everyone who traveled to Saguache for the trial. Many of the men from Union Hill came to view Ellen's case in this special term of the district court. The whole affair ended in an anticlimactic manner when the case against Ellen was nonsuited, meaning

Saguache, with the San Luis Valley in the background. *Author's collection.*

that the judge ruled that Charles Bennett did not have enough evidence to press charges. Ellen and her lawyer returned to Gunnison on the Saturday afternoon stage.

Another incident shows Ellen's ready use of a gun when she felt threatened. Two old-time prospectors who had a "Damon and Pythias" friendship, Tom Riley (or Reilly) and Dick Tullar, were on their way to Chance one night. Tom Riley seldom talked to a woman unless forced into it by circumstances, and here, he had the misfortune of tangling with the notorious Captain Jack. Ellen had pitched her tent in the middle of the trail, and the two prospectors didn't see it as they walked along the trail in the dark. Tom became entangled in the ropes that held up the tent. Captain Jack awoke and let out a volley of curses and bullets. Both men ran away as quickly as possible, with Tom apologizing in his Irish brogue, "Oi didn't mane to do it, missus, Oi tell you, Oi didn't mane to do it!"

In another incident, Ellen used her gun to establish appropriate relationship boundaries. Bill Turner had helped her move some items to her camp south of town, probably to her Mineral Hill camp. The hour was growing late, and Ellen offered him a spot in her tent. Just to clarify the nature of the offer, she placed her gun in the center of the bedroll, saying, "This is your side of the bed, and this is mine, and this [indicating the gun] is to see that you stay on your side!"

In the summer of 1894, confidence in the worth of the Green Mountain Mining District had shot up with two important sales of mines within a mile of Captain Jack's claims. On behalf of Denver investors, William S. Ward agreed to buy the Lucky Strike Lode Mining Claim for $20,000. Ward had obtained his education as a mining engineer from Columbia University, and he had held many prominent positions in Colorado, including curating the Colorado state exhibit at the 1893 Chicago World's Fair. Ward brought much-needed investment dollars to develop the mine, and it also signaled to others that a leading mining man thought the Green Mountain District was worth the investment. The details of the agreement spelled out a cautious approach to investment in the mine. The agreement set out an initial payment of only $1,000, with other payments scheduled over the following year. Perhaps this schedule would allow Ward to see what the mine was worth before shelling out a large sum of money to buy the mine.

Ward expanded his mine holdings in the area by leasing a half interest in the Only Chance Mine from the Gilbert brothers for $10,000. With a reported $1 million worth of ore in sight, the mine had shown promising returns on the carload of ore that had been shipped, and Ward increased production. George Fleming and Alex Craig each owned a quarter interest in the Only Chance Mine, but they were not willing to sell at that price. George's wife, Anne, commented on the stinginess of the Gilbert brothers after their big sale: "The Gilberts…had almost lived at our house, and Mama had cooked endless meals for them; but even though they know how hard up we are, they never treat one of us—never even give one of the children a dime. This was unheard of in mining camps; when one made a sale, he treated."

Anne mentioned that she knew Ellen Jack in Gunnison, and that "we called her Cap. Jack." Ellen told Anne that she had hidden money and government bonds in her bustle when she first came to Denver. A bustle was popular in ladies' fashion, and it consisted of extra padding in the back of a skirt that kept it from dragging on the ground. Anne wanted a bustle, too, so she made one from an empty tomato can that she put under her skirt.

With the promising discoveries of the previous year, legal fights ensued over the ownership of potentially valuable properties. Ellen Jack's former co-owner of the Black Queen Mine George Farnham had to pursue a lawsuit to obtain his interest in the Iron Cap Mine in Spencer. He had a contract with Henry Knight and D.R. Peck at the time that Knight located the mine, which should have entitled Farnham to a one-third interest in the mine. The mine became one of Spencer's leading producers. Farnham won his legal case. Farnham's victory turned out to be hollow, because his lawyers, Alexander Gullet, Sam Crump and Henry Lake, charged him $2,100 for their legal services, and to pay his lawyers, Farnham "sold" them his one-third interest in the mine for the same amount of money. The "courtroom miners" were the only winners in this case. Over the summer, Farnham worked his other mining claims near Ohio City in the northeastern portion of the Gold Belt.

Captain Jack continued her mining activities in the Gunnison Gold Belt by purchasing the Pierson Lode in April. The mining camps of Chance and Iris emerged from their winter hibernation in May, with around one thousand prospectors once again scouring the Green Mountain District for new prospects. By this time, both towns had a post office, general stores, hotels, livery and feed stables, blacksmith shops, assayers and mining engineers, surveyors and several well-built cabins.

Mrs. Captain Jack after a day's work. *Author's collection.*

In Gunnison, Ellen Jack and Emily Rainbow were driving a horse and carriage one day when Ellen lost control of the horse.

"Runaway!" a bystander cried as they careened through the street. At the corner of Main Street and Tomichi Avenue, one block from Jack's Cabin, the carriage overturned, throwing Ellen and Emily onto the ground.

Passersby and businessmen rushed out to help the women. Ellen was out cold.

"Give her air!" somebody yelled.

As some good Samaritan was loosening the clothing around her neck, Ellen quickly regained consciousness and let out "a stream of invective to curl a cat's hair," according to eyewitness J.J. Miller, the young son of the butcher J.D. Miller. Another person described her as "swearing like a bull whacker."

Ellen explained to Doc Shores that she kept a bag of diamonds strung around her neck, and she was afraid that someone was trying to rob her. Doc dispersed the crowd and took Ellen home. Ellen sustained numerous bruises to her face and body, while Emily escaped with only minor bruises.

Emily Rainbow was the twenty-year-old daughter of Winter Stone Rainbow and Mary Ann Rainbow, who were both close to Ellen's age. They lived a couple of blocks from Ellen Jack's residence in Gunnison. The Rainbows were born in England and immigrated to the United States around 1880.

The minor accident didn't stop Ellen for long, as she continued to prospect in the Gunnison Gold Belt. Ellen believed that she had a bonanza when she staked another promising gold claim, the Bright Star. Located a few miles east of Chance in the Cochetopa District, the Bright Star was close to the rich deposit of the Maple Leaf Mine. Ellen reported "some most remarkable pannings of gold," and if the good ore continued with depth, she expected to employ men in sinking a shaft as quickly as possible. Ellen Jack was briefly mentioned by the *Daily Mining Record* in Denver as "doing her share towards the development of the camp," along with many other people who were working their claims and establishing businesses in the new towns.

Located close to Ellen's Scorpion group of claims, Daniel Lehan and Elmer Turner already had a stamp mill at their Mineral Hill Mines, from which they were receiving good returns. They had kept two shifts of miners working all winter. The hard work of developing their mine paid off in May, when they sold their Mineral Hill Mines for the reported sum of $60,000, with $5,000 down, to Oliver P. Posey, one of the leading mining operators in Colorado and the general manager of the Tomboy Mine near Telluride.

The *Saguache Crescent* opined that "when old mining men pay $60,000 for forty-foot holes, you can bet there is something in the ground....The Gold Belt will be a better camp than Cripple Creek inside of four years."

The actual amount of money that exchanged hands appeared to differ from newspaper accounts. The Saguache County records indicate that in November, Lehan and Turner sold the Mineral Hill properties to the prominent lawyer Charles H. Toll for $5,000, with no mention of any later payments or the larger sum of $60,000. Oliver Posey proceeded to develop the property by employing many carpenters to construct a shaft house, a boardinghouse and an office. Three shifts of men sank the shaft to a depth of 175 feet and drove 1,000 feet of drifts. The mine employed sixty-five men, and it supplied ore for a twenty-stamp mill.

In contrast to the boosterish Gunnison newspapers, Anne Fleming conveyed a grittier view of her husband's struggles in trying to scratch gold from the hillsides. At night, George Fleming crushed the gold-bearing quartz in a mortar and used water to pan the gold. This method of recovering the free gold produced a lot of excitement with the early discoveries, but this slow and laborious process could not produce enough colors to make a living. George and his co-owners had shipped two carloads of ore from the mine's forty-foot shaft, but as Anne described, "They take out ore all the time, and it is good, but not enough to pay." The financial hardship took a toll on their family life, and Anne found it hard to keep George in a good mood.

In July, Ellen Jack and John De Nyse located the Mountain Queen claim in Cottonwood Gulch, near Cochetopa Creek. Near this location, Ellen and her mining partners were confronted by a gun-wielding Aaron Duber, often referred to as "Cochetopa Shorty." Shorty accused the prospecting party of cutting his fences, and he objected to them camping on his ranchland. Shorty had some trouble a few years prior with someone who had cut his fences and stolen his goats.

Shorty was well known in Gunnison as a habitual drunkard. He had served in the Civil War, and he collected a pension for his service. Throughout the 1880s, he had been arrested numerous times for public drunkenness. Shorty made one trip into town during which he was not drunk, which was an unusual and newsworthy event reported in the newspaper's social column. By the 1890s, it was forbidden for anyone to sell him liquor, but he always found someone to procure it for him. Shorty was a nuisance and the butt of jokes, but some people looked out for him, too, such as Sheriff Shores when he sent his deputy to bring him to the hospital in Gunnison.

John De Nyse, Ellen Jack and several other men charged Shorty with being insane. They filed a complaint against him for "handling a gun a little too promiscuously" and threatening to "do up those parties who were camped on his premises prospecting." In his defense, Shorty said he had no intention of shooting anyone and that his gun was not loaded. He made the counter-charge that John De Nyse had pulled a gun on him. Shorty was acquitted at his trial, because the jury ruled that there was no foundation for the charge of insanity.

Shorty's life ended a few years later when he accidentally shot himself in the neck and head with his shotgun. He was buried by the Grand Army of the Republic, a fraternal organization that served Civil War veterans. The newspaper described him as an eccentric man. Shorty would be remembered as a "picturesque prospector" in at least two celebrations of the pioneer days that occurred in the 1930s. In the semicentennial anniversary of the town's founding in 1930, the person who represented "Cochetopa Shorty" rode in a buckboard at the front of the parade of pioneers, brandishing a big brown bottle. The "emperor of the Cochetopa" was finally allowed to drink in Gunnison again.

An assessment of production at the one-year mark for the Green Mountain Mining District showed a promising beginning for the new enterprises. An estimated 650 men had worked in the Gunnison Gold Belt in the previous year. A total of over 1,500 tons of ore had been milled and shipped for a value of nearly $50,000. The *Saguache Crescent* argued that if more investment dollars had been available to develop the mines, the output would have been ten times greater. But in comparison to Cripple Creek's output of over $7 million in the previous year, the Gunnison Gold Belt had a way to go before it could be regarded as the next Cripple Creek.

In February 1896, the *Gunnison Tribune* published a rumor that Captain Jack had refused an offer of $40,000 for her claims that adjoined the Mineral Hill properties. The nearby Buzzard Mine had just been sold for an undisclosed "good round sum." With the sale of Lehan and Turner's Mineral Hill Mines for a reported $60,000 to Posey in the previous year and the substantial development of those mines, perhaps Ellen was holding out for even more money. When should someone sell a claim for several thousand dollars, and when should they hold out for millions? Every prospector answered that question differently, and no one could foresee what the future would hold. Would someone sell their claim for a pittance, only for the next owners to make millions? Or would selling a claim for a few thousand dollars be a good return, considering that other owners could sink additional amounts of money into the mine and never see a profit? Some owners would hold onto their claims, refusing good offers, only to see their ore pinch out.

In May 1896, Captain Jack went out prospecting in a new area nearly forty miles southeast of Gunnison, high up in the Cochetopa Hills, where she staked the Hillside Gem claim on the west side of Cochetopa Pass. Ellen located her claim near the promising Silver Plume Mine, which had been discovered the previous summer by the Perry brothers. They had sunk a ninety-foot shaft, and they took out ore that ran high in silver values, along with some gold. Ellen staked another mining claim with W.S. Rainbow and Emily Rainbow that they named Emmy.

The Gunnison Gold Belt received an important visit from Thomas Tonge, the Colorado correspondent for the *Mining Journal* in London, England, and Professor Arthur Lakes from the Colorado School of Mines. They issued favorable reports, urging other investors to get in on the ground floor of this promising mining area. Up to this point, many claims had been located by local prospectors and miners who had been thrown out of work by the silver crash in 1893. They had sunk ten-foot shafts on quartz veins and crushed the ore by hand to pan the gold, but some of the miners didn't even have the money to have an assay performed, and most lacked the capital for further developments. Many of these properties could be bought for reasonable sums of money, in contrast to the outrageous sums demanded for many Cripple Creek prospects; $1 million invested in the Gunnison Gold Belt "would yield handsome profits to the investor."

In fact, Oliver Posey and his associates had acquired the Mineral Hill Mines the year before, and they now expected profits of well over $500,000. With Captain Jack's Scorpion group located just east of the Mineral Hill Mines, maybe she had found a lucrative property as well. Agreeing with Tonge and Lakes, Posey stated that many other properties within several miles of the Mineral Hill Mine could show similar returns with proper development. As visual evidence to support these statements, at their office in Gunnison, the Mineral Hills Gold Mining Company displayed a three-hundred-pound chunk of ore with gold on the surface of it.

Several prospectors made decent sums of money by selling their mines during the summer. Dick Tullar and Thomas Riley, along with two other men, sold their mine for $10,000. In mid-July, John Roberts and J.T. Phillips

Cochetopa Pass, circa 1910. *Author's collection.*

sold the Gunnison Lode for $20,000. They had previously turned down an offer for $10,000, so holding out for more money worked out well for them.

George Fleming and Alex Craig lost a lawsuit against adjoining claim holders. Anne Fleming said that while Sam Crump was a good lawyer, she wished she could have hired Frank Gowdy, who had defended Ellen in a case in the early days of Gunnison. In her autobiography, Anne quoted a long passage from Ellen's autobiography that describes a court case in which Gowdy defended Ellen. The loss of the court case did not stop mining operations at the Only Chance. The *Saguache Crescent* described the Only Chance as "one of the best mines in the Gold Belt." Despite the optimistic reports of high-grade ore, Anne Fleming dejectedly, yet truthfully, stated that even these developments did not produce much gold. The Flemings' living conditions deteriorated. They crammed into a two-room cabin with another family. As baby Neita slept in her bed on the floor, a rat bit her hand, which drew blood and caused her to scream. The Flemings gave up this cabin after the other family moved back to town, and they took up residence in a tiny cabin that had been used by cattle as a shelter.

In October, Captain Jack reported very satisfactory returns for the rich gold ore shipped from her Scorpion group of claims near Iris. She staked several more claims: the Juniper No. 2 and Juniper No. 3, located south of Iris, and the Hillside Jem No. 2, near Cochetopa Pass. She started driving a tunnel into the mountainside and had cabins built so that work could continue over the winter. The snow was deep, and the winter was cold and long at the high elevation of this location among the fir and spruce trees.

The spring of 1897 brought renewed activity to the Gunnison Gold Belt, but luck ran out at Chance for the Flemings. The men had borrowed money the previous year for the stamp mill, and Anne described this investment as "the beginning of the end." The mill broke down, and George mismanaged the mine. Five miners who had been paid only half of their wages won a court case against Fleming and Craig for back wages. As the mine owners were unable to pay their debts, the bank foreclosed on the mine and mill in October. One of the most promising mines in the area had gone broke. In 1898, the sheriff's sale of the Mineral Hill property dealt another death blow to the Gunnison Gold Belt.

George and Anne Fleming moved to Victor, near Cripple Creek. While working at the Vindicator Mine, George drilled into a missed shot, and the explosion killed him. He was thirty-six years old. The *Morning Times* spared no detail in describing his corpse in the morgue: "In the left side of the skull is a gaping, ragged, hideous hole, and the brain, which should be in the skull, lies, a bloody and shapeless mass, beside him in the coffin." Anne Fleming returned to Bonanza, a small mining town in Saguache County, where she became reacquainted with Herbert Ellis, whom she had known in Chance. They were married in 1901. Anne Ellis published her autobiography, *The Life of an Ordinary Woman*, in 1929, as well as two other autobiographical books: *Plain Anne Ellis: More about the Life of an Ordinary Woman* and *Sunshine Preferred: The Philosophy of an Ordinary Woman*.

Captain Jack had found some hidden treasure in the Gunnison Gold Belt but never enough to develop a paying mine. If the rumor of an offer of $40,000 was true, she should have sold her claim.

Ellen Jack went to Denver for health reasons in March 1899. She renewed her legal pursuit of John Kinkaid. Ellen still had not collected all the money owed to her from the sale of the Black Queen Mine. The lawsuit she had filed against John Kinkaid had languished for five years, and in April 1899, she accused him of trying to defraud her.

Kinkaid replied that he did not owe her any money. He stated that Ellen had directed him to hold $7,000 from the sale so that she could avoid paying a court judgment against her that had been obtained by Byron Shear and Isaac Johnson, her former lawyers. Despite her instructions to the contrary,

Kinkaid paid off the judgment to Shear and Johnson. Kinkaid also stated that Ellen was not his client because William Elliott owned the half interest in the mine, although he questioned whether "any such person ever existed."

Ellen had not authorized the payment to Shear and Johnson. She admitted that she had sold the Black Queen to her brother because Shear and Johnson were trying to defraud her. She never quite admitted to making up William Elliott. The closest she came to a confession was when she disclosed that "it was understood that the interest [in the Black Queen] was hers." It is unlikely that her actual brother in England named William Elliott ever came to the United States.

In the April 1899 article that described these legal tussles, the *Rocky Mountain News* began with the byline "Mining Queen in Court," and the story described Ellen as being "celebrated throughout the southwestern part of the state for her enterprise and daring. She is quite wealthy, having made thousands from her mines, which she works herself, and is said to be an excellent huntswoman."

Ellen returned to Gunnison by the summer, where she fought with her neighbor Mrs. Dora Biebel. The two women had been on good terms at one time, as indicated by a poem that Ellen most likely wrote in memory of Dora's brother-in-law, Ernest Biebel, who died in 1895 at the age of forty-nine. He had been committed to the insane asylum at Pueblo because he suffered from being "partially deranged," and he died from "paralysis of the brain." The *Gunnison Tribune* published a poem signed: "A pioneer—E.E.J."

The relationship between Ellen and Dora had soured by the summer of 1899. Against Ellen's wishes, Mrs. Dora Biebel had erected and maintained a chicken coop and stable on Ellen's property for several months. Ellen complained that the chickens were destroying her garden, and the manure pile was causing a nuisance. The situation came to a head one day in June. Ellen threw rocks at Dora, and Dora retaliated by striking Ellen with a hoe, which inflicted a six-inch cut on her scalp. The case was heard by Justice Rogers in September. He dismissed the charges against Mrs. Biebel because Ellen had provoked the attack. Ellen filed a civil suit for damages of one hundred dollars, but the jury ruled in favor of Dora Biebel.

⌒

Ellen's daughter, Adeline, grew up to display many aspects of her mother's temperament. Adeline's marriage to Charles Snevily became rocky when Adeline spent a few weeks with Frederick Holt, a traveling salesman and a married man himself. When she returned, Charles took her back and forgave her, hoping that his wife would change her ways. But one night, Charles came home and found out that Frederick Holt had been there. According to Charles, Holt had an argument with Adeline and beat her. Charles filed charges against Holt. The police encouraged Mr. Snevily to let the matter pass, but he insisted on a court case.

Curious society members wanted front-row seats to see how this drama would play out as they packed the police courtroom in the Town Hall of Flatbush. Adeline entered the arena, dressed in a stunning satin gown, with diamonds adorning her fingers. Frederick Holt bowed to Adeline, and she returned the gesture. Just as Ellen's golden locks had charmed many admirers, Adeline's blonde hair and her flair for the dramatic attracted much attention.

Adeline's story was at odds with her husband's account of events. She testified about the argument she had with Holt.

> We had some words about some private papers of his….I had promised to send them to him, and my failure to do so caused a quarrel. A dozen china cups and saucers which Mr. Holt had given to me were broken. When we were both mad, I said to him, "The best thing you can do is break everything you gave me." He took up one of the cups, and in doing so, loosened the rack. It fell to the floor, and all the cups and saucers were smashed. Then I jumped upon the sofa, pulled down a picture he had given me and tore it up.

Adeline also explained how she had received her bruises. "Mr. Snevily and I had a very bitter quarrel at the breakfast table about some flowers sent to me Easter Sunday. In a fit of rage, Mr. Snevily got up, grabbed me by the throat, pulled me by my hair and struck me. I told Mr. Snevily I considered him a coward to strike a woman. I now have bruises on my arms and body that Mr. Snevily inflicted last Saturday night."

The judge asked Mr. Holt if he had assaulted Adeline, and Holt replied that he had not struck her. The judge dismissed the charges. Adeline left the courtroom with Frederick Holt.

Charles Snevily defended himself in an interview with a reporter: "I am going to sue for divorce, and I think I have enough evidence to get one. I have ninety-six affidavits that tell of her relations with that man and a pile of letters written to her by him. I knew the case would go against me, as those servants and Mrs. Snevily told me they would rather go to jail than testify against Holt. But I'll fix them yet, and don't you forget it."

This salacious story spread far and wide, appearing in the *New York Times*, the *Boston Globe* and the *Chicago Tribune*.

Charles and Adeline managed to patch up their marriage following this public scene. They moved to a new house nearby, but before too long, the neighbors could clearly hear the harsh words in their escalating arguments. At the end of one such fight that their neighbors for a block on either side of their residence could not ignore, Adeline stormed away with Charles's horse and buckboard. The story percolated among the people in Flatbush for a week before it was published in the newspaper in August 1897. Charles was much more reticent about talking to reporters this time, saying that "this is a matter that concerns me alone and a subject about which I do not care to speak." Adeline had left their eleven-year-old son, Charles Jr., with his father, so Charles Sr. placed an advertisement for a woman to take care of his son. The marriage did not survive long afterward.

Adeline, who now went by the name Virginia, married Fred C. Morford, a liveryman. Fred Morford died in May 1902 at the age of thirty-three. Adeline wasted no time in getting remarried in 1902 to Lambert R. Walker, a telegrapher who worked at a stock brokerage. The couple lived in Brooklyn with their children, Edna and William Morford. Virginia and Lambert would remain married until Lambert's death in 1928.

7

THE MINING QUEEN
OF THE ROCKIES

Ellen Jack rented out her residence in Gunnison and went prospecting near Cripple Creek. The *Aspen Daily Times* of February 1900 described Ellen as a "good rustler" who "will make a strike in that camp." The rush to Cripple Creek had begun in 1891–92, and by 1900, the mining district had produced nearly $59 million in gold. The population of the Cripple Creek Mining District stood at fifty thousand, which included Cripple Creek as well as the nearby towns of Victor, Goldfield and several smaller communities.

Ellen Jack ran a boardinghouse at Fourth Street and Bennett Avenue. She had two male boarders, a tailor and a waiter. The Cripple Creek City Directory of 1900 lists her as the proprietor of the Cambridge House, a rather illustrious name for the few rooms located above the J.M. Watts Mercantile Company. Ellen may have gone to Cripple Creek to prospect for gold, but there are no records of any mining claims in her name. In March 1901, she suffered from a severe bout of pneumonia and pleurisy. Her former mining partner Emily Rainbow traveled from Gunnison to care for her for a month.

Ellen Jack won her case against John Kinkaid in March 1904, fifteen years after she had sold the Black Queen Mine. The court awarded her $13,599, which included interest on the original amount. Ellen signed over 20 percent of the award to her lawyer John H. Reddin. It is not known how much of this award Ellen received. It seems unlikely that she ever received very much, because she soon got into trouble when she borrowed money from friends.

Bennett Avenue, Cripple Creek. Ellen Jack's boardinghouse was in the upper rooms of the building on the far left of the postcard. *Author's collection.*

Just as John Kinkaid profited greatly from facilitating the sale of the Black Queen Mine from Captain Jack to the Crystal River Mining Company, he also took part in a large-scale transfer of timberlands from supposedly individual homesteaders to wealthy men. John Kinkaid moved to Idaho, where he won a state senate seat in 1900. He became friends with Governor Frank Steunenberg, and the two men became involved in acquiring timberlands through questionable methods. Steunenberg's life was cut short by an assassin's bomb.

In 1907, John Kinkaid was arrested on charges of conspiracy to defraud the government, along with several other men, including the U.S. senator from Idaho William Borah. The government accused the men of fraudulently obtaining 40,000 acres of land valued at $1 million for the Barber Lumber company. The Timber and Stone Act of 1878 allowed individuals to acquire 160 acres of land for their own personal use by paying $2.50 an acre. Through the Timber and Stone Act, like the Homestead Act, the federal government meant to distribute land to a wide range of the population, an ideal that dates to the founding of the country. Thomas Jefferson believed that westward expansion would act as a safety valve, relieving the tension between labor and capital. Any worker who was unhappy with their position in life could move west and claim free land.

It was legal for individuals to file a homestead claim and then sell their land to a lumber company, and this had happened. But it was fraud if the lumber company paid individuals to file homestead claims for the purpose of selling the claims to the company. These actions undermined the ideal of widely shared opportunity and prosperity and instead resulted in the concentration of wealth. It was difficult for the state to prove these fraud charges. Borah was the first defendant to face a jury trial, and he was acquitted. Two years later, the charges against John Kinkaid and several others were dismissed.

The image of an independent prospector, with a burro, pick and shovel, seeking opportunity and fortune and possibly striking it rich forms a part of the myth of mining. The reality was more complicated. Finding and keeping a fortune happened to only a few people, and much of the wealth from mining ended up in the hands of a few men. Captain Jack drew upon the myth in calling herself the "mining queen of the Rockies." Parts of her life story fit the myth: she prospected, she bought an interest in the Black Queen Mine and she sold it for a profit. She left out or minimized the parts that didn't quite fit: the lawyers who hounded her, the sheriff who threatened to sell the mine and John Kinkaid, the lawyer she once trusted with her own scheme, who most likely kept a good share of the Black Queen's proceeds for himself.

People create their own personal mythos of their lives that give them purpose and fit in with larger cultural myths. The personal myth that guided Ellen Jack's life was that she was destined to find hidden treasure, which fit within the American cultural myth of the quest for gold and silver in the West. The turn of the century also marked a turning point in Ellen Jack's life. She would continue her search by staking mining claims in the North Cheyenne Mining District, located in the foothills west of Colorado Springs. Even more important than her prospecting in this area, she would create and promote the persona of the "mining queen of the Rockies" that would result in her long-lasting fame.

General William Palmer founded Colorado Springs in 1871 as a resort town that catered to wealthy East Coast and British elites. Palmer located Colorado Springs at the base of Pike's Peak, where the prairie meets the mountains. In the early days of the town, about one-fifth of the population was British, earning it the nickname "Little London." Perhaps Ellen felt at home in Little London, but she would never run in the town's elite circles.

The Cripple Creek gold rush had caused a boom in Colorado Springs in the 1890s. The old money elites had to make room for the new millionaires produced by Cripple Creek's gold. The North End of Colorado Springs expanded with the homes of mine investors, mine owners, bankers, lawyers and business owners. Wood Avenue became known as "Millionaires Row." Eugene P. Shove resided on this street, having made his fortune in the Elkton Mine. He was also a director in the El Paso National Bank and a member of the brokerage firm of Shove, Aldrich and Company. Shove had been a cashier at the First Bank of Gunnison in the 1880s, and he was one of the men who had incorporated the Crystal River Mining Company.

Another former Gunnison resident, Samuel Crump, struck it rich in several ways in Cripple Creek. He invested money in leasing the W.P.H. Mine, in which miners discovered a vein of nearly pure gold in 1904. The mine produced $750,000 in two years. He served as a lawyer for the Mine Owners Association. He raked in $10,000 as a special prosecutor of labor leaders during the war between mine owners and labor from 1903 to 1905. The *Rocky Mountain News* criticized the excessive payment, stating that his services were rendered to the Mine Owners Association but were paid for by the State of Colorado. Crump bought a two-story red brick house in Denver for $6,500.

Another Cripple Creek millionaire was Winfield Scott Stratton, a carpenter and prospector who had discovered the Independence Mine. Stratton used his wealth to contribute to the common good in Colorado Springs in several ways. He bought land for the creation of Stratton Park, located near the junction of North and South Cheyenne Canyons. The park had beautiful landscaping and a band shell for concerts, and it was accessible from the city by trolley. Stratton also donated land to the city for the building of city hall. He could have lived anywhere, but he chose to live in a modest home near downtown Colorado Springs. His extreme wealth also brought him troubles. Many people hounded him for money and filed spurious lawsuits against him. When a prospector once asked him about whether he should sell his claim or hang on to it in hopes of getting millions, Stratton answered, "If you get a chance to sell your property for $100,000, do it. I once gave an option on the Independence, and a thousand times, I have wished that the holder had taken it up. Too much money is not good for any man. I have too much, and it is not good for me. A hundred thousand dollars is as much money as the man of ordinary intelligence can take care of. Large wealth has been the ruin of many a young man." Following Stratton's death in 1902, his will provided funds for the Myron Stratton Home for Orphans and Old People.

By 1903, Ellen was residing in Colorado Springs, just three blocks from where W.S. Stratton had lived. In January 1904, Ellen Jack and her mining partner Miss Anna Lustick purchased the Experimental mining claim from the original locators. Ellen and Anna each held a half interest in the claim. Anna mortgaged her property to obtain the money for the Experimental claim, taking a risk on a mining claim that had not produced anything of value yet. She had previously been involved in buying mining stocks, which could also be a risky venture in which someone could lose all they had invested.

The thirty-seven-year-old Anna was born in Iowa, and her parents were from Bohemia. Anna had worked as a cook and servant for Mrs. George Rex Buckman for the past four years. Mrs. Buckman actively participated in the city's leading social events. Her husband had led the chamber of commerce in the 1890s, and he authored the book *Colorado Springs and its Scenic Environs*, which advertised the beauty of the area in order to attract more families from the East Coast and even some from England.

Ellen and Anna's mining claims were located above North Cheyenne Canyon along a scenic route called the High Drive that ran from Colorado Springs to Colorado City. Within the previous year, General Palmer had the road built up Bear Creek Canyon from Colorado City to connect with North Cheyenne Canyon, thus making a loop back into Colorado Springs. From downtown Colorado Springs, Ellen could ride the trolley four miles to Stratton Park and then walk up North Cheyenne Canyon another three miles or so. There had been some excitement in 1891–92, when several men had staked placer claims in Bear Creek Canyon, but no big discoveries had been made yet.

Adjoining the Experimental claim, Ellen also located three other claims: the Mars No. 1, Mars No. 2 and Mars No. 3. She regarded herself as a daughter of Mars, born under that sign, which symbolizes strife and war. Her life had been characterized by plenty of fighting so far, and perhaps that was an apt name for mining claims that she would battle over for the rest of her life.

Ellen hired several men to develop the claims. Paul Phelps, age twenty-five, had worked in gold and silver mines in several western states. On Captain Jack's claims, he first sank a seventy-six-foot incline shaft. The waste rock ran onto the High Drive, and General Palmer stopped Mrs. Jack from dumping the rock on his newly constructed road. Phelps then worked on a tunnel that ran under the High Drive. In 1904–5, Ellen had men drive the tunnel into the mountain for sixty-five feet to look for a mineral vein. Phelps stated that Mrs. Jack "got excited when she ran into a bunch of bull quartz." This type of quartz does not contain anything of value. Phelps stated that "we took pride in the work we was doing for Mrs.

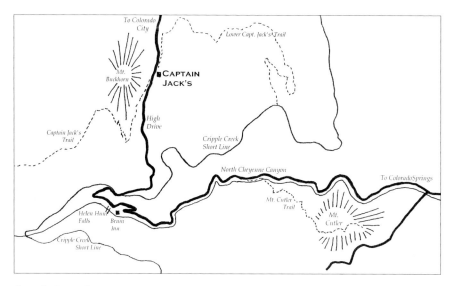

Captain Jack's Camp on the High Drive. *Drawn by the author.*

Jack, because she was a woman, and everyone around there was saying we were trying to get money from her for work we didn't do, so we took pride in what we were doing and did A No. 1 work."

The *Aspen Daily Times* quoted Ellen's description of her claim: "I first found the dyke. It is of porphyry and 20 feet wide. The contact is granite. Two veins were exposed in surface work. They showed gold and copper veins averaging $21 to the ton." Ellen described the mineral veins on her claims this way:

> *There is a mineral chimney up there on the peak, and every vein there is in that hill makes for the chimney. It is just like the spokes of a wagon wheel; and the veins all make for the chimney, and they shoot in every direction from this point right opposite to my house, and you can't go through that hill and develop one claim, but what you must develop all that is on that hillside, because they are shot in every way from this mineral chimney; from this tremendous blow-out that is up there.*

Ellen Jack reported a promising start to ore production from her prospects. Assays on the ore showed returns ranging from $20 to $260 per ton. Ellen shipped a ton of ore to the local mill, and she anticipated hauling large quantities once the nearby Cripple Creek Short Line built a spur to Camp Jack.

Mrs. Captain Jack's camp, with Paul Phelps. *Author's collection.*

Mrs. Captain Jack with Paul Phelps. *Author's collection.*

This shaft was more for show than mineral production. *Author's collection.*

Ellen engaged in hyperbole with this claim. Only a small amount of gold appeared in assay reports. *Author's collection.*

On Monday, December 19, 1904, Captain Jack found a bundle of burned clothes in the woods as she walked toward town from her place on the High Drive. She wondered who had burned their clothes in this spot. Upon her arrival in town, she found out that a woman had been murdered on Cutler Mountain. She notified Sheriff Grimes and Police Chief Reynolds of the clothes she had found. She also reported a brief incident that she thought might be related to the woman's murder. In late November, Ellen had been approached by a German man who asked for directions and acted strangely.

The murdered woman had been discovered two days earlier on Saturday afternoon by three men while they were drawing property boundary lines on Cutler Mountain. The woman's nude body lay facedown over a log. Her face was burned beyond recognition. An autopsy revealed that she had died from a .38-caliber gunshot to the head. A dentist described expensive dental work that could be used to identify her.

This unsolved murder case appeared in newspapers across the country. Ellen Jack's story of finding the clothes made it into several papers, such as the *Rocky Mountain News*, the *San Francisco Call* and the *Salt Lake Herald*. The *Washington Times* described Ellen as a medium and spiritualist. The paper quoted her description of a dream she had about the murder:

> *I saw in my dream a man of slight build, and apparently between thirty and thirty-five years of age, lead a young woman to the top of the mountain, evidently on the pretext of showing her a mine.…There was another man on the spot, a German, hiding behind a tree. As the young man and woman stopped here for a moment on their way down the slope, the German fired from behind the tree. As the girl fell, the young man bent over her with clinched fists and said: "Now, you will give me away, will you?" But the woman was dead. The young man and the German hurriedly stripped her of her clothing, built a pile of stumps, laid her across it, then took a bottle and poured the contents over her head. A fire was lighted, and I seemed to see the flames shoot upward.*

Ellen seemed to have incorporated many of the facts that were known about the case. She also included her own story about the German who had acted strangely. A man working for Ellen stated that the German she had seen was a harmless old Dutchman who asked about the way to Clough's camp. The Dutchman went on his way to this camp and was later found to be working there.

On December 26, there was a break in the case when a Colorado Springs hairdresser recognized the woman's hair color and texture as that of Bessie Bouton. Her family contacted Police Chief Reynolds and told him that they suspected Milton Franklin of murdering their daughter. Originally from Syracuse, New York, Bouton had met Franklin in New York City. She earned a decent living as a sales agent for beauty products, traveling in many states and Canada, and eventually selling her products in Denver and Colorado Springs. At first, her family liked Franklin, but they grew distrustful of him after they discovered that Franklin was a gambler and pool shark. Bessie had told her sister that she knew something about Franklin that could put him in prison.

Franklin's running from the law ended almost a year later in San Francisco. Four policemen went to investigate a report of a suspicious woman. A police officer posed as a plumber to gain entrance to the apartment. The woman let him in, and he made a show of taking some measurements as he tried to see if Franklin was hiding somewhere. When the officer stepped out of the room for a just a moment, the woman quickly slammed the door behind him and locked it. Two gunshots rang out in the room. The officer shot the lock off the door, and all four policemen entered the room. Franklin's companion Nulda Petrie had fallen backward onto the bed with a bullet to her brain, and Franklin had collapsed on the floor after he shot himself in the temple. The police discovered a statement written by Franklin in which he denied any involvement in Bessie Bouton's death. The police suspected that he had also murdered three other women.

Captain Jack said that she continued to hear the death shriek of Bessie Bouton echoing through the canyons on still nights at her place on the High Drive.

It didn't take long for Ellen to become embroiled in conflict. In November 1904, she charged another prospector in the Cheyenne Mountain area, Victor McLind, with threatening to take her life. He countered that she had threatened him, and he filed a complaint against her. On December 1, Ellen was acquitted of the charges in Justice McClelland's courtroom.

It is unlikely that Ellen ever received any more money from John Kinkaid for the sale of the Black Queen Mine. She should have received an additional $10,800 after winning her court case, which would have given her plenty of money to invest in her new mining ventures. Instead, she borrowed money from her mining partners Anna Lustick and Josephine Smith.

Ellen borrowed $150 from Josephine E. Smith in October 1904. Mrs. Smith worked as a cook for Ellen from December 1904 to April 1905, feeding the miners whom Ellen was boarding at her claim on the High Drive. About ten years younger than Ellen, Josephine was born in Bohemia and immigrated to the United States in 1865. Josephine had recently obtained a divorce from her husband, Charles C. Smith.

The marriage had been contentious for many years before the divorce was finalized in June 1904. Josephine charged Charles with extreme cruelty, and she described an incident in which he "struck, beat, choked and bruised" her, with further injuries prevented only by the intervention of their sixteen-year-old son. Josephine stated that Charles refused to support the family, even though he worked as a tinsmith. She had to work as a dressmaker and washerwoman to bring in money. Charles deserted the family, and he failed to appear at the divorce trial. A jury of six men ruled in Josephine's favor and awarded her alimony, to be paid for by a lien on the house. From 1902 to 1904, Josephine Smith lived about three miles south of downtown. Mrs. Smith said that she took out a mortgage on this house to loan Ellen Jack the money for a patent application for her mineral claims. In 1905, both Ellen Jack and Josephine Smith lived in the same residence near downtown.

Ellen applied for a patent to her gold-bearing claims: the Mars No. 1, Mars No. 2, Mars No. 3 and the Experimental Lode. This patent application started troubles between Ellen and her two friends. Ellen didn't have much money, so she proposed a scheme to Anna Lustick. Ellen convinced Anna to sign over her half interest so that it would be easier to get a patent. Once Ellen got the patent, Anna's half interest would revert to her. In addition, Ellen borrowed $200 from Anna in December 1904, and she gave her a promissory note, stating that she would pay her back within five months. Anna said that Ellen told her that "she owned the Black Queen Mine in Gunnison, and that it was being sold in Denver, and she would pay me back double. Under that promise, I loaned her the money." Ellen may have been expecting further payments from the judgment she had won against Kinkaid, which she could have used to repay Anna.

Ellen filed the patent application in June 1905, but she still needed to do additional work for it to be complete. Ellen borrowed an additional $368 from Anna, and she spent the funds on wages for the men who worked on her claims. Believing that she owned a valuable mine, she intended to pay back the loan when the mine produced high-grade ore.

Anna had given Ellen nearly all the money she had from her job, which paid her $35 to $40 per month, plus an additional $300 Anna had received from her brother. One time, when Ellen was sick, Anna asked her for a promissory note for the additional borrowed money. In response, Anna said that Ellen "jumped

An early photograph of Captain Jack's camp. The two women may be Josephine Smith and Anna Lustick. *Author's collection.*

off of the cot, and she told me she didn't owe me anything." Anna also found out that Ellen no longer owned the Black Queen Mine. Anna stated, "If she had been square with me, I would have said, 'Mrs. Jack, I am willing to wait until you can pay me.'" But instead, Ellen denied owing her money.

In January 1906, Anna filed suit against Ellen to recover the money. In Ellen Jack's cross-complaint, she stated that she had invested $800 of her own money into developing the Experimental claim and that as a half-owner in the claim, Anna owed her $400 for those expenses. Anna won a judgment for $616.

Anna's brother Frank Lustick had loaned money to Ellen, and he also filed a complaint against her in January 1906. Ellen's understanding of the agreement was that she would pay back the loan with proceeds from the "production, if any, from said mining property, in other words, an investment to be repaid if results warranted." Ellen believed that Frank had brought this suit prematurely, as she had not found ore *yet*. Frank won a judgment against Ellen for $400.

Ellen had a falling out with Josephine Smith as well. Ellen and Josephine lived at the same address in 1905, but by 1907, Josephine was renting a room nearby. In addition to money troubles with Anna and Frank, Ellen couldn't repay the money she had borrowed from Josephine Smith either.

Josephine Smith retaliated against Ellen by taking her mining claim. Ellen had secured her promissory note to Smith by placing her Mars No. 1 mining claim with the public trustee of El Paso County. Smith requested that the public trustee sell the mining claim. The sale took place at the county courthouse in July 1907, and Smith was the high bidder at the sale. Josephine Smith now owned the Mars No. 1 claim.

Josephine Smith tried to get Ellen in trouble over her widow's pension. In December 1907, Josephine wrote a letter to the commissioner of pensions in Washington, D.C. She asked, "Can a woman, a widow, who marries again after her husband's death, draw a pension?" Commissioner Warner wrote back to Smith. He explained that Ellen Jack's pension had been restored because the Gunnison County Court had declared her marriage to Walsh null and void because Walsh was already married to another woman. Ellen retained her pension.

Ellen's biggest discovery on the High Drive was that she could make money from selling her persona as the "mining queen of the Rockies" to the tourists. She had been living in a large canvas wall tent before she built a twelve-by-fourteen-foot cottage that provided better protection from the cold weather. From the side of the cottage, she served meals and refreshments in a rustic, open-air patio with tables, chairs and white tablecloths. She conveyed the image of herself as the "mining queen of the Rockies" in a series of self-published postcards. Postcards came into existence in 1901 with a change in postal regulations, and they quickly became very popular. Postcards of major tourist attractions in the Colorado Springs area included those

The Bruin Inn in North Cheyenne Canyon on the way to Captain Jack's Camp. *Author's collection.*

from the Bruin Inn, Helen Hunt Falls, rock formations in North Cheyenne Canyon, Seven Falls in South Cheyenne Canyon, Manitou, Pikes Peak and Garden of the Gods. Many postcards also depicted large and small mining operations, such as the mines in Cripple Creek. It was unusual for a woman to make postcards of herself and then sell them to tourists.

One person wrote the following description on the back of one of the postcards: "Home of Capt. Jack, oldest living member of the Eastern Stars, lunched here Aug. 3, 1912, elevation 9,500 ft. Since her husband's death, she has taken the name of Capt. Jack. Is the only woman at the cabin. Prepares and serves lunch every day." The Order of the Eastern Star is a women's group affiliated with the Masons. Ellen's late husband had been a Mason. Ellen published a poem on a postcard that honored the Masonic fraternity. The heading in the poem referred to Charles Jack's service under the command of Flag Officer Farragut when Union forces took control of New Orleans and Baton Rouge.

Lines Written by Mrs. Ellen E. Jack,
Widow of Capt. Charles E. Jack, of the Farragut Fleet, U.S. Navy

MASONIC WELCOME HERE

Welcome to our Colorado, brothers from both far and near,
To see our mountains, valleys, and streams, and the glorious sunrise that is
to be seen from here.
And the morning star arising in the East, so beautiful and fair,
And the Shriners with their bright red caps and the Royal Arches with their
dazzling regalia
Make our mountains ring with mirth, as they come to have a jolly good
time here.

And the hands they grasp and say: This is the hand of fellowship,
But not a hand to slay, except some foe should his teeth show and try to tear
our craft asunder,
Then the sword they would draw and with a roar as though a peal of
thunder sound from shore to shore;
Then the Sir Knights, with the Eastern Stars so bright, would gather from
both far and near,
And with a prayer to their lord God to give them victory
And save them from being rent asunder.

On another postcard, the sender wrote, "Here is a card I bought from Cheyenne Mt. Mrs. Jack keeps the inn at the highest point of the High Drive [on] Mt. Cheyenne during the tourist season and spends the balance of the year prospecting as you see. She is 65 years old and a talented woman, having composed a GAR [Grand Army of the Republic] song."

Several tourists mentioned Captain Jack in their travel logs that were printed in their hometown newspapers. One visitor from Illinois called Ellen a "unique mountain character….She is seventy years of age, owns and runs a mine of tungsten ore, keeps her shack open for light refreshments and sees nearly everyone who comes." Two of Ellen's postcards mention tungsten ore. But there are no other mining reports of any tungsten deposits near her mining claims.

Captain Jack kept a couple of burros at her roadhouse and photographed herself with them. The hardworking burro had once been extensively employed in various aspects of mining: pulling ore carts in the mines; carrying supplies over narrow, steep and treacherous trails; and hauling ore from remote mines to towns or railroad shipping points. Mechanization replaced the burro's usefulness in mining, and many of the animals were turned loose to roam the hills, forage for garbage in towns or serve as children's pets. Burro rides became a staple of the tourist trade in Colorado

Mrs. Captain Jack, the mining queen going to her tungsten mine. *Author's collection.*

There are not any known tungsten deposits in this area. *Author's collection.*

For the grandest trip in the Rocky Mt. Region and No Toll!

TRAIL TO MT. CUTLER

Take a burro or pony at the

Rustic Burro Stand

(On the right)

For the trip to MT. CUTLER, over the O'Brien Trail viewing both Canons and Seven Falls.

ROUND TRIP 50c.

H. T. O'BRIEN, Manager.

Exclusive transportation privilege to Mt. Cutler.
Saddle Horses for Bruin Inn and Palmer High Drive.
Competent guides with each party.
Take Stratton Park, Cheyenne Canon Car or Carriage.

O'Brien's Balloon Trip to Pikes Peak and return (in one day) on horses.

Phone Main 907. (Over)

Tourists taking burro rides. *Author's collection.*

Springs and Manitou. Wilbur O'Brien took tourists on burro rides all over the area, including up through North Cheyenne Canyon and over the High Drive to Captain Jack's place, where the tourists looked at her mine tunnel.

Another colorful Colorado Springs character, Nora Gaines, took tourists on carriage rides over the High Drive, past Captain Jack's cabins and down Bear Creek Canyon into Manitou. "Ma" Gaines, as she was also called, had a rugged appearance. She wore men's clothing and had a heavy build, and she handled her horse and buggy team as well as any man and swore alongside the best of them. Like Captain Jack, she was a tough character who could hold her own in a man's business.

Mrs. Gaines picked up tourists at the Denver and Rio Grande train depot, competing for customers alongside the other carriage drivers. Many of the male drivers harassed her because they resented what they perceived as her intrusion into their trade. In one such incident, driver Burton Townsend was fined twenty dollars for using insulting and abusive language toward her. No shrinking violet herself, Nora Gaines swore at Police Officer McReynolds when he tried to get her to pay her hack license, and she was arrested for creating a disturbance.

On one carriage ride, Mrs. Gaines saved her party of tourists from a robber. They had descended from the High Drive, and just outside of Colorado City, a highwayman stepped into the road. He brandished a revolver and ordered Gaines to stop. Instead of submitting to the man's commands, she whipped her horses and sped ahead. The would-be robber grabbed a woman passenger's sealskin coat; he ripped off the coat's arm and fell to the ground. He landed under the carriage, and the rear wheels ran over his legs. The tourists had already checked out of the Antlers Hotel that morning, and on their return to town, they took the first train to Denver on their way back home to San Francisco.

Some tourists selected Gaines because it was unusual for a woman to be a carriage driver. Tourists recommended her to their friends and family. By 1913, Gaines no longer had a monopoly on being the only woman carriage driver. She got into a heated verbal battle with Mrs. Boyd when both were trying to attract a party of tourists. Fierce competition among drivers often led to verbal sparring. Boyd charged Gaines with disturbing the peace, but a jury found her not guilty. "Mother" Gaines also had a generous side to her character. She gave gifts of pies and fruitcake to the Colorado Springs police officers at Christmastime. Her late husband, Daniel Gaines, had been a member of the force. She also sold Christmas wreaths that she made from evergreens and kinnikinnick.

In *To Colorado's Restless Ghosts*, authors Hunt and Draper describe Captain Jack and Ma Gaines as "friends of a sort, but the kind of friends who could enjoy each other but without any of the amenities. They fought like tigers, but neither cared for the opinions of others or for personal appearance. Each could hold her own. Each had a healthy respect for the other, and there was no competitive feeling." Hunt and Draper also reported that Captain Jack had commented that she "regretted the necessity for ladies like us to have to associate with the likes of Nora Gaines."

One of Ellen's newspaper advertisements read: "Don't fail to go over the High Drive. Fine chicken dinner served in home style at the summit of the High Drive for 50 cents each." A visitor's directory in the *Colorado Springs Gazette* read, "The High Drive: one of the most picturesque mountain highways in the world. An exclusive carriage ride through beautiful North Cheyenne Canyon, stopping at the Bruin Inn and Captain Jack's unique refreshment pavilions." From 1907 to 1909, Ellen was charged with selling liquor without a license. Maybe that was what the advertisement meant by "unique refreshment."

Captain Jack had run saloons in Colorado's pioneer days, but now, she faced pressure to close her roadhouse on the High Drive. General William Palmer had banned saloons in Colorado Springs by including a clause in property deeds that prohibited the selling of intoxicating liquors. He didn't have any objection to people of his social standing drinking alcohol, and

Ellen encountered a mountain lion while taking meat to miners at the Black Queen. She said, "You go your way, and I will go mine." *Author's collection.*

Mrs. Captain Jack's refreshment pavilion. *Author's collection.*

people of the right social class could go to the numerous drugstores that filled prescriptions of alcohol for "medicinal" purposes. He wanted to prevent the working-class saloon crowd from setting up shop and creating problems in his town. The saloon and red-light district flourished two miles to the west in neighboring Colorado City.

Woman's Christian Temperance Union president Frances Willard organized the Colorado Springs Union in 1880, just six years after the founding of the national organization in 1874. In addition to speaking against the evils of alcohol, the WCTU had a Department of Mercy, which battled against animal cruelty and family violence, as well as a department devoted to social purity that tried to suppress prostitution. By the turn of the century, Colorado had ninety unions in towns throughout the state and a membership of 1,600, with the largest union in Colorado Springs.

Carry Nation toured through several Colorado cities in the summer of 1906. Mrs. Nation had been the president of a Kansas chapter of the WCTU when she started her saloon-smashing crusade in 1900. She claimed to hear the voice of God directing her to "Carry A. Nation" to prohibition, which some people took as evidence of insanity. Mrs. Nation lectured in Colorado Springs in August. She promised to smash a storefront the next day, and a crowd gathered at the Owl Drugstore to witness her hatchet-wielding stunt. Instead, she merely broke a bottle of beer with her hatchet. She called out the hypocrisy of Colorado Springs residents who had banned saloons but allowed alcohol to be sold as "medicine" in the drugstores, telling the crowd, "You don't know the meaning of prohibition; you are all a lot of drunkards and cigarette fiends." Carry Nation's crusade was part of a larger temperance movement that brought about changes in laws and customs in the United States regarding alcohol.

Captain Jack's roadhouse came under scrutiny as local authorities cracked down on several saloon and brothel owners. Ellen Jack pleaded not guilty to the charge of selling liquor without a license in January 1907. A twelve-man jury heard the case, but by the next day, they were unable to agree on a verdict. Over the next seven months, the court summoned Ellen to appear an additional five times to face these charges, and each time, the case would be reset for a new trial date.

As Captain Jack evaded the liquor charges against her over the summer months, the beer glasses continued to clink together amid the merry laughter of revelers at her roadhouse on the High Drive. Two lovers rode on horseback up North Cheyenne Canyon to her roadhouse for their trysts. The nineteen-year-old Laura Mathews had recently come from Chicago to Colorado Springs to seek relief from her "nervous troubles," accompanied by her nurse Tillie Green. Laura had met Amos Rumbaugh in Chicago, and he pursued her to Colorado Springs. Rumbaugh told people that he worked for the government, possibly the secret service, but he evaded questions about the exact nature of his mission in Colorado Springs. He had abandoned his wife and child in Pennsylvania to have a good time in Colorado. He drank a lot and had plenty of cash to throw around. Amos bragged that while many men pursued the beautiful Laura Mathews, she went out on carriage rides with only him.

One night in July at Captain Jack's, Amos asked Laura to marry him, and he flashed $1,000 worth of bills in front of her. She spurned his advances. Laura seemed despondent about all the money Rumbaugh was trying to lavish on her. Ellen advised her to take the money, if for no other reason than to keep him from spending it foolishly. Ellen thought that Laura was afraid of him. Ellen also suspected that the nurse Tillie Green had hypnotized Laura. Ellen said that "the strange look on the girl's face when the party was at my place told me she was in trouble." Ellen felt that something dark was about to happen—perhaps someone might try to rob Rumbaugh of his roll of bills or worse. As Laura and Amos left her roadhouse that night, Ellen rode with them to the mouth of the canyon, taking her pistol for protection. Laura's parting words to Captain Jack were, "I'm going on a long journey, and you'll not see me again."

Laura met her death just two days later. Laura's body was found on Broadmoor Hill, along the road, lying in the mud, her outstretched arm pointing to the gun that lay two feet away. The initial ruling was death by suicide from a bullet to her brain, but many people suspected she had been murdered. The news did not surprise Captain Jack.

Rumbaugh did not show up at the coroner's inquest two days later. The undersheriff went to look for him. Tillie Green testified that Laura had been despondent for some time, and she expected that Laura would die by suicide. As Miss Green was testifying, the undersheriff returned from his search for Rumbaugh. He flung open a nearby window into the courtroom and shouted, "Rumbaugh has shot himself!"

A hotel clerk had entered Rumbaugh's room and found him in the bed with a bullet wound to his temple and "his brains oozing out." Barely hanging on to life, Rumbaugh was taken to a nearby hospital, where he died the following morning. The coroner's juries ruled both deaths were the results of suicide. People wondered what secrets Rumbaugh took to his grave. Had he murdered Laura after she rejected his offer of marriage and then taken his own life?

Captain Jack thought she had some useful clues to the mysteries that surrounded the case. She went to Sheriff Grimes and shared her suspicion that Tillie Green had hypnotized Laura. She also told him that Rumbaugh had over $1,000 when he was at the roadhouse just days before his death. Only $300 was found on his person at the hotel.

In giving newspaper interviews about the case, Captain Jack also made sure to describe her past as a "hero of frontier days." She told a story about fighting Natives, and she explained that the scar on her forehead came from a tomahawk blow. This newspaper account appears to be the first time this false tale appeared in print. She had acquired the scar in a bar fight on Christmas Day over two decades ago in Gunnison. She also recounted the Civil War legacy of her husband, who had taken part in a daring mission to cut the chains of the Confederate blockade, which led to Admiral Farragut's capture of New Orleans.

Tillie Green behaved in a cold-blooded manner that aroused suspicions, but the sheriff did not find evidence of her involvement in any crime. The coroner's jury rulings of suicide stood firm, and many mysteries about these events remained unsolved. Laura's spirit continued to hover around the roadhouse on the High Drive, and Captain Jack would still hear the young woman's pleading voice whispering through the pines.

Perhaps the wild times on the High Drive and the death of Laura Matthews contributed to the attempt to close Captain Jack's roadhouse. One temperance lecturer at the Methodist Episcopal church and the Baptist church described the saloon as a "murderer, in that it not only slays the better self of the drinker but that the traffic incites deeds of violence."

Ellen Jack won the first round of sparring over the liquor cases when Assistant District Attorney Henry Trowbridge announced in late September that there would be no further prosecution. But Ellen would face these charges again the following summer. In August 1908, the court charged her with four counts of illegal liquor selling. She faced additional charges of keeping a saloon open on Sundays and selling liquor without a county commissioner's license. She pleaded not guilty to all charges. She believed that the law interfered with her rights, and she vowed to fight the charges once again.

General Palmer backed the liquor charges against Ellen. Around this time, Ellen wrote a flattering poem about the Colorado Spring's founder. The *Chicago Tribune* published this poem in 1911. Ellen also wrote "The Veteran's Cry."

The crackdown on Captain Jack's roadhouse continued when Sheriff Grimes seized her beer. Due to the judgment Anna Lustick had obtained

LINES WRITTEN BY MRS. ELLEN E. JACK,

Widow of Capt. Charles E. Jack, of the Farragut Fleet,
U. S. Navy, in Honor of General William J. Palmer's
Gift of tne High Drive to the City of
Colorado Springs.

"Extras! Extras!" was the dismal sound
When they fired on Fort Sumpter
And tore Old Glory down.
A fair, slim man sprang to his feet and cried No! No!
They had better go slow, for William Palmer with his
  regiment will go
And strike those traitors down!
And they fought 'till the enemy was badly beaten
And Old Glory waved all over land and peaks.

Then to the mountains General Palmer came,
With an eye for beauty, and the wild west scenes,
And a city he founded they call the Springs,
Whose fame for beauty spread everywhere 'till the
Tourists came from far and near.
And its climate in winter so mild, dry and warm,
And in summer the cool mountain breeze;
And the High Drive he built for the tourists to go
Is the grandest on earth, for they all say so.

Now in the year nineteen hundred eight
To the city General Palmer did donate,
And forty thousand to keep the road clear,
So the tourists would have no fear
Of driving up to the summit there.
And on the top of this High Drive
They will hear the hack drivers say, "See yonder!
Is Captain Jack's widow standing by the way;
She lives all alone and knows no fear, but welcomes
  all who come up here.
And shows them all the scenery, so grand,
And bids them adieu, with a shake of the hand.

Tho' in the lonely twilight hours
A passerby might hear her say,
Since thou art gone I have wandered here
And my golden locks have turned to gray.
Tho' the dust of years is on your grave,
Who fell to make this country free,
Thou art gone, I will fill your place.
And in honor of this noble gift
I will raise this flag upon this hill,
And watch Old Glory wave on still.

(Author of The Fate of the Fairy, or 27 Years in
the Far West, and the song, "The Veteran's Cry.")

A poem honoring General William Palmer. *Author's collection.*

LINES WRITTEN BY MRS. ELLEN E. JACK,
Widow of Capt. Charles E. Jack, of the Farragut Fleet,
U. S. Navy.

P. O. Box 453, Colorado Springs, Colo.

The Veteran's Cry

At my camp fire, sad and lonely,
Backward turn my thoughts tonight,
On that story of Old Glory and on the
Noble boys that fought, yes, they fought
To save Old Glory with a soul of strength and might.
And to show to all Nations,
Let them be black, brown or white.

Tho' the veterans' forms are bent
And their hair is snowy white,
We permit no dictation and no insult
To the Stars and Stripes;
For we have left our sons and daughters
To be ready at the Nation's call,
And in trouble they will be ready
To start along with the bugle's call;
For the veteran's cry will ever be
Oh, God! protect our flag and keep this nation free.

Ah, what do we hear! The veteran's cry so near,
Sons and daughters true, we leave that work for you
    to do—
That not a rent, or tear shall mar
That flag that laid us here.
Tho' it be party, creed, or crowned head,
It matters not, for the veteran's cry will ever be
Oh, God! protect our flag and keep this nation free.

[Author of the book, "The Fated Fairy," "The
Indian Girl's Prayer" and "The Lesson That You
Taught."]

A poem honoring veterans. *Author's collection.*

against Ellen in 1906, Sheriff Grimes placed a levy on her property, which included her mining claims and her supply of beer. The sheriff took possession of three barrels and numerous quarts of Budweiser, Blue Ribbon and Goetz beer. In January 1909, at a public auction at the county courthouse, Sheriff Grimes sold the beer to the Phillips Smith Drug Company for twelve dollars. Perhaps the drugstore could now sell the beer for "medicinal" purposes. This small amount of money did not even cover the sheriff's expenses. A few weeks later in February, District Attorney Ferguson stated that there would be no further prosecution of liquor cases against Ellen.

Ellen did not mention these cases in her autobiography, but she did give her opinion of temperance reformers, stating: "I find a lot of people banded together as temperance unions who have been reformed drunkards and had to take a big oath and have a mob at their backs to keep them from getting drunk. These are all degenerates, for if man or woman cannot drink beer, wine or liquor without getting beastly intoxicated, he has no willpower and is a degenerate. These people are a curse to the country....They want to take the rights of the level-headed people away."

Captain Jack continued to operate her roadhouse on the High Drive. In 1909, she said that she found a cave that she wanted to turn into a tourist attraction. The nearby Cave of the Winds attracted many visitors, so it was plausible that there might be another similar cave in the area. She kept the exact location of her find a secret because she was looking into acquiring the

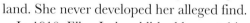

land. She never developed her alleged find.

The cover of Ellen Jack's autobiography. *M.A. Donohue & Co., 1910.*

In 1910, Ellen Jack published her autobiography, *The Fate of a Fairy, or Twenty Seven Years in the Far West.* She closed her autobiography with a rambling rant in which refers to herself as a fairy: "And after twenty-seven years of wandering through the Rocky Mountains in the Far West, I drop amongst the people as though I came from the clouds.... Hark, all ye Hebrews; hear the prophecy of the fairy who wears a crown of silvery gray; hear, now, the time is coming fast when this persecution [against the Jews] shall cease."

Ellen Jack did not explain why she called herself a fairy. She may have been drawing from Victorians' fascination with fairies. Science made many advances in the latter half of the nineteenth century, including Darwin's publication of *Origin of Species* and *Descent of Man*. Cultural anthropologists sought to make their contribution to science by gathering fairy tales and seeking their origins and evolutions. One school of thought stated that fairy tales originally described prehistoric races of small people and that the stories had been passed down from that time. Some people regarded fairy tales as quaint stories of a fantasy land, while others regarded fairies as real beings.

Some people thought that fairies continued to exist in the modern world. Sir Arthur Conan Doyle, the creator of the detective Sherlock Holmes, believed that science would eventually discover the reality of fairies, and he made his case in *The Coming of the Fairies*. He stated that fairies were as "numerous as the human race" and that they were "separated from ourselves by some difference of vibrations." Just as there were parts of the color spectrum not visible to the naked eye, it was possible that fairies existed in some realm that would be uncovered by scientific investigation. He gathered stories from many people who reported seeing fairies. It was possible that these people could see fairies that others could not see.

Almost half of Ellen's autobiography discussed her days before coming to Colorado—her birth and childhood in England and her marriage to Captain Jack. She described traveling to Denver, then on to Gunnison and finding the Black Queen Mine. The book contains several photographs of her as a miner that were taken near her place on the High Drive. The book played an instrumental role in creating the oft-repeated legend of Captain Jack, as people would continue to reiterate the stories in various newspaper and magazine articles. Western Reflections Publishing Company republished her autobiography in 2010 with the title *Colorado's Eccentric Captain Jack*.

One notable area of fabrication and contradiction was what Ellen wrote about Natives. She most likely made up her story of fighting Natives in Gunnison. A tourist who met Ellen and sent one of her postcards in 1909 wrote on the back, "She had a tomahawk scar upon her forehead since an encounter with Indians." There are no independent records of her killing anyone, although she made plenty of armed threats against men who were trying to cheat her.

Why would she have made up this story? Recalling one's past takes place within a social and cultural context that can affect how people tell stories. People sometimes acknowledge that they are not always completely accurate in telling stories and that they may alter details to be more entertaining. Captain Jack told her stories to entertain the tourists when

"This is the gun that Captain Jack killed the Indian with." *Author's collection.*

they visited her place on the High Drive, so it's quite likely that she added a few exciting details that fit the script of what might have happened to a pioneer in the Wild West.

While Captain Jack's tale of fighting and killing Natives might offend the modern reader, it fit into the cultural context of the early 1900s in the way that white settlers memorialized and glamorized the pioneer days. For example, *Buffalo Bill's Wild West* had been reenacting scenes of fighting and killing Natives in its popular shows since 1883. William Cody had been a scout for the army in the Indian Wars, and he had killed Yellow Hair and other Natives. In his shows, he declared a change in his views, stating, "An enemy in '76, a friend in '85." He employed a great many Natives in his shows; he paid them well and maintained friendships with many of his performers for years. Historian Louis Warren describes Cody's attitude toward Natives as "ambivalent." On the one hand, Cody acknowledged that Native hostilities resulted from miners' and settlers' invasions of Native land, yet he also profited from the opening of the Black Hills to miners, as well as from his Wild West shows.

Some white people criticized and lamented the nation's treatment of the Natives, although this viewpoint was not as common during this era. Colorado Springs author Helen Hunt Jackson published *A Century of Dishonor*, in which she chronicled numerous abuses against Natives, such as broken treaties, stolen land and massacres. She argued for stopping these abuses and correcting these wrongs committed against Natives. Ellen conveyed these views in a poem:

The Indian Girl's Prayer.

Who! Is this great God? They are all talking about,
The Indian Bee girl is trying to find out,
But with the paleface books and the knowledge they claim
Not a soul ever lived that ever found out,
What this mighty power in its glorious Shrine.
The poor Indian girl goes to the glorious Sunrise and with
Upturned face she cries, to the great spirit who gives us
These Mountains, Valleys and Streams and freedom to roam
Around them like the Birds and the Bees.
With no creed to control her, no prison to enslave her,
With the cruelty of your so-called civilized laws,
Whilst some are enthroned in Mansions with millions,
And they turn away from the poor starven papooses
With contempt and scorn.
They call us heathens, savages, and tell us that their Jesus died for us
To be saved. But this is bad pay for the land they have stolen
And driven us away.
Our camp fires are all out, our Wigwams all gone,
With no Mountains to roam,
To the glorious Sunrise to worship the great spirit.
That controls Sun, Moon and Stars and all above and below,
Which gives us the instinct the way we should go,
With freedom to all, injustice to none,
Then our prayers will be said and our songs will be sung.

While the tourist trade brought in some money, Ellen wanted to further develop her mining claims, but in doing so, she became entangled in bitter fights with her former friends Anna Lustick and Josephine Smith. Ellen tried to regain the Mars No. 1 claim from Josephine Smith by filing a lawsuit in 1911. Ellen insisted that she owned the claim, and she charged Smith with trespassing on her property. Mrs. Smith defended herself by stating that she owned the claim. Mrs. Smith charged Ellen with jumping the claim and threatening to kill the men who worked for her. The jury found in favor of Josephine Smith. Ellen tried to gain control of her other mining claims by filing a patent application in July 1911 for the Mars No. 2, Mars No. 3 and Experimental claims. Anna maintained that those properties belonged to her.

"Give it up? Never!" Ellen vowed to fight "until the last ditch." She believed that her mining claims contained vast wealth. In fact, there had not been any recent reports of her receiving payment for mineral production. Ellen also believed that she could develop her roadhouse into one of the fine resorts that catered to tourists in the area.

Ellen's daughter, Adeline, who was married to the wealthy stockbroker Lambert R. Walker, wanted Ellen to live with her in New York City. Ellen rejected her daughter's offer of the "luxuries of a Fifth Avenue drawing room." Why would Captain Jack want to sit, stiff-backed, in fine furniture, hemmed in by floor-to-ceiling windows and surrounded by walls adorned with fine art when she could sit comfortably on her rustic pine open-air porch with nature's canvas at her doorstep? Captain Jack had spent far too long in her independent lifestyle in frontier Colorado to settle down to a life of luxury back east, where social conventions would have been like a straitjacket compared to the freedom of the hills. Captain Jack did not want to leave behind her two old cats who kept her company on the High Drive, even at the price of not being able to see her grandchildren. She was as "happy as a lark" at her picturesque home amid the pines.

W.F. Conway of the *Rocky Mountain News* described Captain Jack as "the epitome of the sublime and the ridiculous…a character that is as much a part of the High Drive as the road itself.…On the night air, she sends out a diapason from her old melodeon and sits and gazes at the stars. Anon she communes with the elements, a modern Viking riding the fierce storms. Messages flash from the skies, she says. All is clear to the mind which delves in the occult." An admission of getting messages from the skies might have been her ticket to an asylum if she had resided back east.

Ellen described another reason for preferring her mountain home to New York City. She wrote, "I look farther, and I find they have built railroads in New York and built buildings twenty and thirty stories high. In every city where there are large sheets of water, there is a liability of a volcanic eruption which they call earthquake. When they come, I would rather be in the mountains than near those high buildings."

The case of *Anna Lustick v. Ellen Jack* went to trial in November 1912, in the District Court of El Paso County, with Judge J.W. Sheafor presiding and a jury of six men hearing the case. The main dispute centered on who

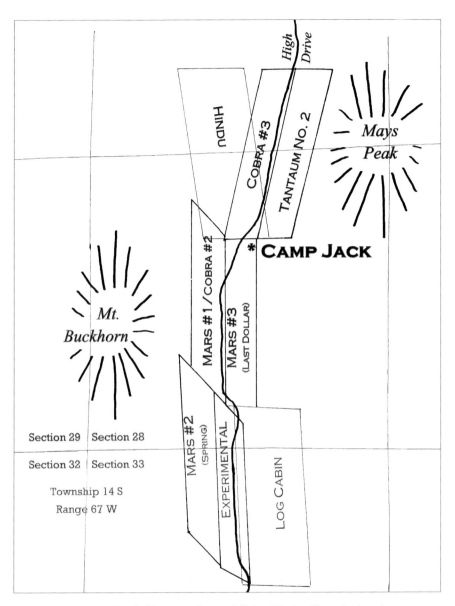

Mining claims in the North Cheyenne Canyon Mining District. *Drawn by the author.*

held the legitimate claim to the mining properties in the North Cheyenne Mining District.

On the surface, the women fought over possession of mining claims, but underneath those legal issues, the case exposed in open court the personal animosities that had developed between Captain Jack and her two former mining partners. Two months earlier in September, Josephine Smith had leveled serious accusations against Ellen. According to Ellen, several other people were present at her place on the High Drive when Josephine said, "She never was Captain Jack's wife. She is posing here as Captain Jack's widow. She was nothing but his prostitute, nothing but his harlot."

Contentious accusations continued in the courtroom. In a cross-examination of Smith and Lustick, Ellen Jack's lawyer Mr. Orr pursued a line of questioning that implied they were vindictive in trying to take whatever belonged to Ellen. Smith and Lustick stated they were trying to get the money owed to them, but they also revealed the strong feelings they held against Ellen Jack.

Mr. Orr questioned the gray-haired Josephine Smith on the witness stand. Referring to Anna's attempt to gain control over the disputed mining claims, Mr. Orr asked Mrs. Smith, "When she [Anna] gets the three claims she is after now in this lawsuit, you and Miss Lustick will have everything that Mrs. Jack ever had?"

Mrs. Smith replied, "Which we honestly ought to have, sir—an honest debt."

Mr. Orr asked, "That takes everything she has?"

"She took everything I had; I lost my home, I lost my furniture; I had to sell my kitchen range to get my goods up from the hill and live down here. She never paid me a cent. She has her pension, and I have nothing, only my two hands to earn my living with, and I am an old woman, only four years younger than she."

"When did you and Miss Lustick make your arrangements to get all this property?"

"I made arrangements through the court. It was my just due, to satisfy an honest debt which she owed me. Miss Lustick did her own business to suit herself. I have nothing to do with Miss Lustick's part of it. She [Lustick] is an honest woman; that one [pointing to Mrs. Jack] is a thief."

Judge Sheafor intervened, "Mrs. Smith, you will have to control yourself."

Smith replied, "Pardon me, your honor, I will try to."

"You will have to do it, not try to."

"I will then."

The *Rocky Mountain News* reported these courtroom theatrics with the title "Gun Woman Quarrels in Mine Litigation—Captain Ellen Jack Fights Suit So Bitterly She Is Called 'Thief' by Witness."

Mr. Orr questioned Anna Lustick, and after her attorney objected to his questions, Mr. Orr explained himself, saying, "It is, if the court please, for the purpose of showing the covetous mind of this plaintiff [Anna Lustick], beginning way back at that time, to get these properties, and it is perfectly competent and proper."

Mr. Orr continued his questioning of Miss Lustick, asking, "When was it, Miss Lustick, that you made up your mind that if you couldn't get these claims of Mrs. Jack by judgments that you were going to locate them and take them away from her in that way?"

Anna defended herself and dodged the question, saying, "I had a right to do what the law allows me; it was open ground."

Orr continued, "Then when you got your money, why weren't you satisfied to let the old lady keep her property?"

Anna answered, "I want what belongs to me and what the law allows me, thank you." Anna also stated that the legal proceedings provided her with only a portion of the money owed to her.

Orr questioned Anna about locating her claims in 1909 and whether she had seen Captain Jack at her buildings on the High Drive at that time. Anna answered, "I didn't say I never had seen her. The trouble she has put me to, I wish to God I never had seen her. I have been sorry many times I ever met that old woman. She beat me out of everything; that is what she tried to do."

The hostilities continued outside the courtroom as well. According to Ellen, as she approached an area where Josephine and Anna were standing in the presence of several other people, Josephine said, "Here comes the old whore now." According to Ellen, Anna also made accusations against her in the presence of witnesses and jurors during recesses of the court, saying to Ellen, "You are an old thief; you get men drunk on the High Drive and rob them, you dirty old thief."

After four days of testimony, Judge Sheafor issued extensive instructions to the jury. He emphasized that pertaining to the question of whether the mining claims were open to location by Lustick in 1909, the jury had only to consider whether Captain Jack had done the necessary assessment work in 1908. The jury returned a verdict in favor of Anna Lustick for all three mining claims.

Captain Jack wasted no time in filing a motion for a new trial. The court overruled this motion. In response, Captain Jack announced her intention to appeal her case to the Colorado Supreme Court.

Why would Josephine, Anna and Ellen engross themselves in such long and bitter battles over mining claims that had little mineral value? Ellen's previous battles had been over valuable properties—the Black Queen Mine and her claims in the Gunnison Gold Belt. Anna and Josephine had given Ellen what little money they had earned from their low-paying women's work in the hope that they might become a little better off. All the women were poor, yet they all had lawyers to carry on their legal fights. It seems that the reasons were quite personal; Josephine and Anna felt that Ellen had lied to them and that she had betrayed their friendship and trust.

The issue of fraud comes up frequently with mining claims. The *Engineering and Mining Journal* discussed this question and delineated several categories of people:

> *1. The deliberately dishonest: selling stock but not having a mine.*
> *2. Semi-dishonest: acquiring a mine but selling at an inflated price not justified by conditions and not believed in by themselves.*
> *3. The honest, but ignorant: people who deceive themselves and others*
> *4. The visionary: Colonel Sellers.*
>
> *Only the deliberately dishonest break the law, but the others are just as dangerous.*

Colonel Sellers is a character in Mark Twain's novel *The Gilded Age*, which describes the "futility of chasing after riches. This theme is personified in the character of Colonel Sellers, who sees 'millions' in countless visionary schemes, though he rarely rises above grinding poverty," according to Barbara Schmidt in her article "We Will Confiscate His Name: The Unfortunate case of George Escol Sellers." In his autobiography, Mark Twain described the relative who inspired the character of Colonel Sellers:

> *James Lampton floated, all his days, in a tinted mist of magnificent dreams and died at last without seeing one of them realized. I saw him last in 1884.…He had become old and white-headed, but he entered to me in the same old breezy way of his earlier life, and he was all there, yet—not a detail wanting: the happy light in his eye, the abounding hope in his heart, the persuasive tongue, the miracle-breeding imagination—they were all there; and before I could turn around, he was polishing up his Aladdin's lamp and flashing the secret riches of the world before me.*

This description of Colonel Sellers seems to fit Ellen better than the other categories regarding her claims on the High Drive. Or she may have been honest but ignorant of the fact that her claims had little value. She possessed the charm and the perpetual optimism of a prospector, and she convinced Anna and Frank Lustick, as well as Josephine Smith, to invest in these claims. But they charged Ellen with being semi-dishonest, with having a mine that she knew was worthless, thus stealing their money.

Captain Jack also seemed to carry on the fights for a very personal reason—she had to defend her story and legacy as the "Mining Queen of the Rockies." Everyone constructs a myth of their life, which is a coherent story that ties together the separate episodes and forms the basis of one's identity. In writing her autobiography, Ellen described her role as a heroic pioneer, recounting her adventurous life in a positive tone. Josephine and Anna attacked this self-image, which Ellen would protect "until the last ditch." And in the meantime, Ellen's tourist business depended on maintaining this image.

In 1915, Ellen wrote the lines to the following song, and Chas. L. Lewis composed the music. She added a photograph of herself at her cabin on the High Drive on the front of the sheet music.

The lesson that you taught
If you had come to me, sweetheart, many years ago
To ask this question, I should not have answered you with no.
That was a time when I knew naught, but since then I have learned
Of trust betrayed, hope deferred, of love half won, half spurned.
How could I know you only sought, to pass an idle hour
Unmindful of how you pluck'd or dropp'd your once admired flow'r
How could I dream your vows were false since mine were only true?
If I am changed at last, sweetheart, I've learned the lesson taught by you.
And now you come again, sweetheart, tho my head is white as snow,
To ask my love my trust, that has perished in the years ago
And so, farewell I do not hold for you an unkind thought,
Nor yet regret for what has passed, the lesson that you taught,
Of heartlessness, indifference, or call it what you will,
My tender heart so surely learned, and heeds the lesson still.

This song brings up several unanswered questions. Was she talking about her past romantic relationships or a current love interest? Did the bitterness of her fights with Anna and Josephine spill over into this song? Is it autobiographical at all?

The sparring continued between Ellen and Anna for the next couple of years, with the two women trading places as plaintiff and defendant in court cases that appeared on the trial docket. Anna Lustick eventually won her battle to take all of Captain Jack's properties on the High Drive. Anna acquired mineral patents for the Mars No. 2 lode in August 1916, followed by the Last Dollar Lode in January 1917.

In the meantime, Captain Jack refused to give up staying at her cabin on the High Drive. Ellen continued to entertain tourists as they passed by her curio shop. In 1915, one tourist and her party from Altoona, Pennsylvania rode up the High Drive in a carriage, where they were "introduced to the quaintest of mountain characters—Ellen E. Jack, better known as Captain Jack of the mountains, who welcomes all who visit her mountain home and exchanges pleasant banter with all." Ellen was mentioned again in 1916 by another tourist from North Carolina who wrote about his travels in the area.

In January 1917, Ellen was hospitalized for pneumonia. The *Colorado Springs Gazette* described her as "one of the few women prospectors of the early days in the Gunnison and Telluride districts, but of later years proprietor of a refreshment pavilion on the High Drive." Someone took a terrible photograph of Ellen in her sickly condition in the hospital. Dressed in a hospital gown, with her eyes closed, Ellen was propped up by three nurses who were standing next to her. During her sickness, Ellen's two beloved cats were either lost or taken from her cabin on the High Drive. She offered a reward to recover them, but it is not known whether her cats ever returned.

Ellen Jack sued Anna Lustick in November 1917 for $10,000 in damages for spreading rumors about her. Anna Lustick denied the charge that she had called Ellen a thief at the courthouse. Anna's lawyer requested that Judge Sheafor instruct the jury that if they did find evidence that Anna made those statements, they should understand that she made them in the heat of passion because she was provoked by Ellen's behavior and that they should rule in favor of Anna. Judge Sheafor refused to issue those instructions. The judge instructed the jury that should they find in favor of Ellen: "You will, in that case, assess the plaintiff [Ellen Jack] with only such damages you judge the plaintiff's reputation actually suffered. And in that case, you will take into consideration all the circumstances attending the utterance of such words, their character, their falseness, the malice, if any, displayed by defendant [Anna], the provocation, if any, given by the plaintiff [Ellen] and also the rank and position in society of the plaintiff and the defendant." The jury found in favor of Ellen Jack, and they awarded her damages of $1.

Ellen Jack with tourists, 1915. *Author's collection.*

Ellen filed for bankruptcy in January 1918. Anna Lustick appeared before the referee in the case to protest the bankruptcy, but she failed to file the necessary paperwork. The bankruptcy was granted in May, and thus Ellen discharged the $1,400 in debts from previous court judgments that she still owed to Anna Lustick, Frank Lustick and Josephine Smith. The bankruptcy filing also revealed that Ellen had few assets and no money with which to pay her debts. The equity she had in her house on Tejon Street amounted to $800, and the value of her household goods, clothing and stock of goods at her "summer resort" on the High Drive totaled $250.

Left: "Returning from the mine." *Author's collection.*

Below: Ellen Jack with parrots and burros. *Author's collection.*

Opposite: Mrs. Captain Jack with her cats. *Author's collection.*

MRS. CAPT. JACK,
THE MINING QUEEN OF THE ROCKIES

Ellen continued to occupy her home on the High Drive. In April 1919, she traveled to Pueblo to visit William Mosher; his daughter, Iris Mosher; and T.A. Jones. They planned to visit Ellen's homestead southeast of Pueblo. A glowing article in the *Pueblo Chieftain* described her remarkable life, spinning stories both true and false. She owned a ring that was given to her late husband, Charles Jack, by Robert E. Lee (maybe true, but not confirmed by other sources). She described emptying her six-gun while fighting Natives (false). Ellen stated that she sold her Black Queen Mine for $25,000 (true, if one combines the amount of the sale to all three partners). One year later, the mine sold for $3 million, and it had a later valuation of $42 million (probably false, as these latter two figures are not corroborated by other sources).

While Ellen reveled in the glory days of having once owned a valuable mine, she lived in poverty in her later years. Her pension had increased to twenty-five dollars per month in 1917 due to a Congressional act that increased pension rates. In 1919, she wrote to the commissioner of pensions. She believed that she was entitled to a larger pension increase because her husband had been an officer and she was now over seventy-five years old. She did not receive any additional money.

For the last few years of her life, Ellen suffered from poor health. Yet she persisted in occupying her home on the High Drive. In May 1919, she put out an advertisement in the *Colorado Springs Gazette* that stated, "DON'T MISS going over the High Drive, most beautiful scenery in

Mrs. Captain Jack, mining queen of the Rockies. *Author's collection.*

the world. Mrs. Captain Jack is at home there and will be glad to see her many friends again."

In 1920, Ellen moved her buildings from Anna's Last Dollar claim to the adjoining Cobra No. 3 claim, which she still held. The sparring continued when Anna tried to take Ellen's remaining claim from her. Anna had a survey done as part of her application for a mineral patent for the Hindu Lode, which mostly covered the same ground as Ellen's Cobra No. 3 claim. Ellen jabbed back by having a survey completed for the Tantaum No. 2 Lode, which adjoined her Cobra No. 3 claim to the east. A.J. Walker, Ellen's daughter, was listed as the owner of this claim. Ellen won a pyrrhic victory on May 3, 1921, when she received a patent for her Cobra No. 3 claim. She was dead a month later.

The search for hidden treasure had been Ellen's mission in the latter half of her life. She pursued her fortune, which was unusual for women at that time. She drew on Quaker traditions to follow her inner light. Ellen prospected, she purchased and staked mining claims and she overcame obstacles to make a profit with the Black Queen Mine. In the last decade of her life, Mrs. Captain Jack fought for her legacy as the mining queen of the Rockies. And she would win this battle.

8

MRS. CAPTAIN JACK'S LEGACY

llen E. Jack died on June 16, 1921, due to leakage of the heart. She had become despondent after a flood washed out the High Drive and she was unable to return to her mountain sanctuary among the pines. Her grief aggravated her heart condition and contributed to her death. Similar to lifelong spouses who die within a short time of each other, Captain Jack could not survive the loss of being unable to return to her beloved claims and cabin on the High Drive that she had fought so hard to keep in her final years. She said that she would fight "until the last ditch" to keep her place, which she did until the forces of nature prevented her return. Modern research shows that "broken heart syndrome" causes a weakening of the heart and an excessive release of stress hormones, which can cause death in people with a heart condition.

Ellen's funeral was held by the ladies of the Grand Army of the Republic, a fraternal order of Union veterans of the Civil War. She was buried in Evergreen Cemetery. Ellen's obituary in the *Colorado Springs Gazette* described her as "one of the West's most remarkable personages." The *Gunnison News-Champion* described her life as "full of adventure, hardship and courage. The history of this remarkable woman reads like a page from some thrilling romance and all the more interesting because it was true." Both newspapers recounted several aspects of her life, including her being born in England, marrying Charles Jack, arriving in Gunnison's early days, owning the Black Queen Mine and serving tourists at her place on the High Drive.

Ellen E. Jack's tombstone in Evergreen Cemetery. *Jane Bardal, photographer.*

Many obituaries in newspapers around the country described her legendary status as a Native fighter. The *Albuquerque Journal* headline read, "Noted Woman Indian Fighter, 69, Dead," and the article referred to the scar on her forehead from a tomahawk. Similar stories appeared in many newspapers, including the *Salt Lake Telegram*, the *Rocky Mountain News*, the *Denver Post* and the *Colorado Springs Gazette*. The *Gunnison News-Champion* included the claim that she fought Natives in Gunnison's early days, but it admitted that "there is a seamy side to the life of this wild west character.… It is doubtful, for instance, whether her stories of Indian wars had any foundation in fact."

The heading from a brief article in the *New York Times*, "Can't Break Mother's Will," referred to Ellen's legal will, but it could also be read as saying that Captain Jack would get her way this time as well. The article mentioned that Ellen Jack had received $50,000 on her husband's death. Maybe her daughter, Adeline Walker, expected that Ellen still had great wealth. In addition, with Ellen using Adeline's name over the years in property transactions, Adeline might have expected to receive real estate holdings in the will. Adeline traveled from Brooklyn, New York, to Colorado Springs for the funeral and the reading of the will.

Ellen Jack left five dollars to her daughter. Such an action is usually meant to convey the message: "I didn't forget you—I intentionally left you nothing." Ellen had been described as a wealthy woman at several points in her life, but she had declared bankruptcy in 1918, and there would have been little left in her estate to fight over. Captain Jack named two of her longtime friends, William W. Mosher and Alberta J. Walker (no relation), as her executors.

Adeline Walker contested the will. She filed a petition to the court in July 1921. She claimed that the will was invalid because her mother suffered from insanity, illness and old age. She accused Alberta Walker (also known as Roberta) and William Mosher of deceiving her mother and of having "undue, improper, and illegal influence" over her.

Judge Kinney heard the case in September. Clyde Phillips and Clyde Starrett had been witnesses when Ellen Jack signed her will in May 1920. Their testimony provided insight into her mental state. Druggist Clyde Phillips had known Captain Jack for about fifteen years. Attorney Clyde Starrett asked Phillips, "Are you able to say whether she was of sound and disposing mind on that occasion and realized what she was doing?"

Phillips answered, "Oh, yes, I think she realized what she was doing at the time. She seemed about as rational as I had ever seen her, in fact."

"What was her sickness?"

"I think it was asthmatic trouble or something."

"Did you ever, in your acquaintance with her and at the time of the making of this will, know of Roberta Walker having any undue influence over her?"

"No, not to my knowledge."

"In your acquaintance with her, did she ever express any dislike toward her daughter, Adeline?"

"No, I never heard her say anything about that."

C.B. Horn, the attorney for Adeline Walker, cross-examined Clyde Phillips: "Did you ever notice any peculiarities about her?"

Phillips replied, "Yes, sir, I always knew she was peculiar."

"Did you ever notice any peculiarities such as would indicate to your mind that she was not in her right mind?"

"No, I never thought of that."

"Did you ever hear her discuss the fact of having her acts and doings controlled by the planets or anything of that character?

"No, I never heard her say anything about that."

Ellen Jack had stated some of her ideas about planets in her autobiography: "There are three powers that control all things—electricity, vibration, and what I call the Push is an aura light that is around every living thing and controls every creature, and works with planetary vibrations, according to the position of the planets. The God head essence is in this light."

Lawyer Clyde Starrett also testified in the case. Ellen Jack had contacted him about drawing up her will. He was not well acquainted with her but had done business with her several times since 1916. He stated that "from my knowledge of her, she was a very peculiar woman. I always considered her strong-minded. I think *erratic* expresses it about as well as I know how to express her mental condition, but at no time did I ever consider her to be insane."

Attorney Horn cross-examined Starrett: "You say that you never noticed anything peculiar about her that would indicate any mental infirmity, or insanity, or disease of mind?"

Starrett replied, "I hardly know what you mean by mental infirmity. She was different from the ordinary run of human beings in her general ideas about mining, and about religion, and about business, and possibly about ethics, but she always knew what she was doing."

"Have you ever read her books?"

"Yes, sir, read one of them."

"Was there anything in that book that would indicate that she was suffering from any mental infirmity?"

"I have read a great many more that have been more poorly expressed, and I have read books on the market that were much worse written than that."

"Did she ever have any talks with you, Mr. Starrett, about the fact that the planets or stars controlled her actions and conduct?"

"No, I don't think she ever did. She has on one or more occasions indulged in some of her planet talk, and I laughed at her. That was about all. She knew that I didn't have much respect for her theory, and she didn't try to impose it on me."

"I will ask you, Mr. Starrett, if it is not a fact that she was a married woman at the time of the Civil War?"

"Yes, sir, I think so, from what she claimed to me that her husband and General Lee were friends, and I believe that would have to be before the Civil War, because her husband was in the navigation service. I might state, too, that she was a woman of peculiar ways. She had a habit of referring to people by their last name, seldom by their given name. Always called Mrs. Gaines, Gaines, Mr. Mosher, Mosher. It was a trait more pronounced in her than I ever noticed in anyone else. She would call Mr. Penrose, Penrose, Mr. Krause, Krause. She always referred to them by their last name."

Judge Kinney ruled that the will was valid, stating that Ellen Jack had been of sound mind and had not been under any undue influence from anyone else at the time it was made. William Mosher and Alberta Walker carried out Captain Jack's wishes for the disposal of her property. They sold her house at 320 South Tejon Street for $3,000. There was little money left over after paying $2,144 to the bank for its lien and $193 in taxes. Ellen Jack's lot in Gunnison sold for $100 in January 1923.

Captain Jack's personal property brought in only $75 in a sale held in February 1922. No auctioneer would agree to sell her dilapidated furniture. A box full of copies of her book *The Fate of a Fairy* did not even merit an estimate of value. Her copyright was listed as having no value.

Most people would not receive much money for their claims against Ellen's estate. William Mosher claimed $1,368 against Ellen's estate based on a promissory note, but he would receive less than half of that amount. Alberta Walker received the ring that, according to Ellen, Charles Jack had received from General Robert E. Lee. It appraised at $35. She also received a watch and chain valued at $10. Ellen bequeathed the remainder of her jewelry to Mary Ann Rainbow, whom she had known in Gunnison in the 1890s, but there was no record of any additional jewelry to distribute.

Anna Lustick requested $1,500 from Ellen's estate for the value of three buildings that Captain Jack had taken from Lustick and moved to her property, but Anna did not receive any money. Anna Lustick took over some of Captain Jack's mining claims on the High Drive by obtaining mineral patents in the following two years. The mineral survey shows discovery cuts and tunnels on these claims, but there are not any records of Anna doing any mining. As alleged by Captain Jack's attorneys, perhaps Anna merely wanted a victory over Captain Jack. Miss Anna Lustick died in September 1925 at the age of fifty-seven, and she was buried in Ely, Iowa.

Nora Gaines bought Captain Jack's place on the High Drive in 1922. There, she constructed a large building that held a dance hall. Gaines continued to call the place Captain Jack's, and she offered meals and dances to the tourists. She probably ran the place for only a few years. She was stricken with paralysis in 1924, and she died in 1933 in Colorado Springs. The buildings fell into disrepair, and in 1965, the City Parks and Recreation Department razed and burned them because they were too far gone to save as a landmark.

Captain Jack had bequeathed the remainder of her estate to her grandson Charles Snevily of Brooklyn, New York. Nothing remained after assorted bills were paid, and no record stated whether he received any proceeds from her estate. In addition, no record indicates whether her daughter, Adeline Walker, received her five dollars.

The daughter who accused her mother of insanity ended up in a mental institution. (Adeline) Virginia Walker became an inmate of Kings Park State Hospital for the Insane by 1935, and she still lived there at the time of the 1940 and 1950 censuses. Adeline Walker's son Charles Snevily died in 1927. Her husband, Lambert R. Walker, died suddenly from a stroke in 1928. He left his estate to his wife, whom he referred to in his will as A.J. Virginia Walker.

It is not known why Adeline ended up at Kings Park State Hospital, due to the confidentiality of medical records. She most likely was incarcerated due to mental illness, as that was the reason most inmates were there. But the hospital also treated people with tuberculosis, and it had a geriatric department as well. Virginia Walker could have been there for any of those reasons. It is not known how long she stayed there. A Virginia Walker died on July 13, 1964, in New York, according to New York State's Death Index. It is hard to verify if this is the same person as Ellen's daughter, but one clue is that her birthdate is listed as 1871, which is close to Adeline's birthdate of 1869 or 1870 in other records.

Kings Park State Hospital was opened in 1885 as a "lunatic farm" on Long Island to care for the insane and the poor. Originally built to relieve overcrowding in New York's other hospitals, by the 1890s, the Kings County Asylum had over 1,000 patients. Asylums often became warehouses that kept mentally ill people out of sight. At the time of Virginia Walker's residence in 1940, Kings Park State Hospital had over 6,100 inmates. The hospital grounds contained one hundred buildings. Completed in 1941, the brick, twelve-story Building 93 housed 1,200 patients. During this era, doctors experimented with more invasive techniques, such as insulin coma therapy, lobotomy and electroshock therapy. Many doctors who used these techniques viewed them as progress in treating mental illness, because patients became more emotionally calm. Critics charge that the procedures left many patients worse off than before, because the techniques caused brain damage and some patients experienced the treatments as punishment or torture.

Ellen Jack is not known to have been diagnosed with a mental illness, nor did she spend any time in a mental institution, but she certainly was regarded by many people as eccentric or peculiar. In *Eccentrics: A Study of Sanity and Strangeness*, David Weeks and Jamie James described common characteristics of people who considered themselves to be eccentric, starting with the cardinal feature of nonconformity. Ellen clearly fit this characteristic. She went against many societal norms throughout her life: she divorced Walsh and spoke openly about his abuse in a public trial; she owned the Black

Queen Mine and she prospected in numerous locations; she ran a roadhouse mostly by herself on the High Drive.

They claimed many eccentrics also display creativity and intelligence. Ellen used these qualities to get out of several seemingly impossible jams: she created William Elliott, which kept her pursuers at bay at least for a while, and she convinced a jury of men that she was not guilty of pension fraud. Eccentrics are often opinionated and outspoken. Ellen fought against liquor charges and railed against prohibitionists.

Weeks and James distinguish eccentricity from mental illness, although the line between the two is not always clear. Eccentrics might have unusual thought patterns, but they regard their thoughts as a positive and unique aspect of themselves. In contrast, people with schizophrenia often experience distressing delusions or hallucinations. In her autobiography, Ellen described hearing voices that guided her actions, such as when she was about to marry Redmond Walsh and a voice told her, "No!" And when she was traveling to Aspen in a snowstorm, a voice told her, "Go on," which saved her from an avalanche. These voices served as warnings or guides.

Both men and women prospectors were often regarded as eccentric. Sally Zanjani, in *A Mine of Her Own*, described several features of woman prospectors. Even though Ellen Jack was eccentric or unusual in comparison to other women of her era, she shared many features with other women prospectors, many of whom were just as unusual.

Zanjani described two women prospectors who were committed to insane asylums. Mary E. Stewart, also known as Mollie Monroe, prospected near Wickenburg, Arizona, in the mid-1860s. She wore men's clothing, drank whiskey, chewed tobacco and swore "harder than any man in Arizona." Her "sins" were described as "a reckless abandon for legal proprieties in the matter of marital affairs." She would take and then leave a husband as she pleased, without the formality of a marriage or a divorce. For unspecified reasons, she was charged with lunacy and found to be insane in 1877.

After Mary spent about a year and a half in confinement, ex-governor Safford visited her in the asylum. She told him that her "mind was completely restored" and that she wanted her freedom, and she promised to abstain from alcohol if released. Safford concurred with her statements and said that "she could now be safely turned out of the asylum." But another report by a doctor said that while Mary could be rational for weeks at a time, this was not always the case, and at those other times, she required watching and restraints. Mary was not released. She sought her freedom by escaping on a scorching hot day in late April 1895, but the sheriff tracked her footprints in the desert sand and returned her to confinement. She died in the institution in 1902.

Sally Zanjani also wrote about "Happy Days" Diminy, who prospected in the Nevada desert for several decades. At the age of eighty-six, Diminy beat the district attorney over the head with her cane. He attempted to get her committed to the state mental hospital, but a doctor ruled that she was sane. She went to San Francisco one winter to tell fortunes, but "what Nevadans in the deserts around Tule Canyon could tolerate as cantankerousness and even admire as fiery spirit, San Franciscans considered madness." She was committed to the institution at Stockton, and she died there in 1948.

Captain Jack did many of the same things these women did: she drank, she swore, she had multiple husbands, she believed in spiritualism and she was cantankerous. One has to wonder why Ellen escaped the fate that befell these other women.

Captain Jack shared several other characteristics that were typical of women prospectors, as described by Zanjani. Many of them regarded raising children as incompatible with their work. Some placed their children in religious boarding schools or with relatives. Ellen Jack left Adeline with relatives when she headed west. Male prospectors do not seem to have been judged along this standard, probably because men were not expected to take care of children.

Many women prospectors did not stay within the confines of traditional marriage. Some had younger husbands, and some did not marry at all. It is unclear how many times Ellen Jack was married, because in her autobiography, she did not mention being married to several men. According to a court document and other newspaper references, she married Jeff Mickey, and according to rumors, she lived as husband and wife with Frank Royer.

Many women began mining after age forty, as contrasted with men, who were more often mining in their twenties or thirties. Women who started prospecting later in life often did so because they no longer had children to care for or their husbands had died. Ellen Jack traveled to Gunnison at age thirty-seven after her husband and three of her children had died.

A woman traveling alone had to defend herself. "Mountain Charley," a woman who dressed and passed as a man, once had to defend herself with guns when threatened by several men. Ellen Jack had several instances in which she used her guns to defend herself or her mining interests. She stated her philosophy that guns made men and women equal.

Women prospectors often did not follow organized religion, but many had a unique life philosophy. Ellen regarded herself as a Quaker, a spiritualist and a Rosicrucian, and her idiosyncratic views on many topics make sense when viewed as part of these traditions.

Many women prospectors lived long lives, which Zanjani ascribes to their resilience and resourcefulness. The opposites of those qualities, helplessness and hopelessness, contribute to early mortality. Control is a major factor in a person's well-being and longevity. People who have more control in their lives are less depressed and have lower rates of heart disease. Socioeconomic status also has a major impact on health, with wealthier people living healthier and longer lives. Captain Jack had her widow's pension that provided a small but steady income, which allowed her to have more control over her life. She was described as a wealthy woman at several points in her life, although a good share of the proceeds from the Black Queen ended up in her lawyers' pockets, so it is hard to know how much wealth she had at any given point. She declared bankruptcy in her later years, and she did not have any great wealth at the time of her death.

Money and control also affected a woman's ability to choose a husband as she pleased. In an era when most women had to marry for economic survival, having the means to make this choice was important in maintaining control of one's life. Women in abusive marriages have high rates of depression. Captain Jack had the means to divorce Walsh, whom she accused of extreme cruelty.

A love for the outdoors was enjoyed by many women prospectors, whether it was the Arctic, the desert or the mountains that inspired them. Zanjani describes one woman who said that "when she lost the strength to go out in the wilderness, she would prefer to die." Captain Jack had searched for hidden treasure for the latter half of her life, finding it in many ways: making money with the Black Queen Mine, prospecting in several other locations and enjoying the scenery and serenity of her home on the High Drive. Her heart broke for the last time when she could no longer access her mountain retreat.

In 1912, W.F. Conway, writing in the *Rocky Mountain News*, feared that characters such as Captain Jack would soon be forgotten, saying that "some day, not far distant, when the curtain is wrung down on the 'Captain Jacks' that have given the West a distinctiveness of its own, its picturesqueness, or much of that romance and glory associated with the hills, will have gone. Then…will the West have truly lost something that modern civilization with its never-ending progress shall finally cease to recall."

Captain Jack has not been forgotten. She ensured her place in history by writing her autobiography and publishing photographs of herself. In the 1950s, Betty Wallace interviewed old-timers in Gunnison who still remembered colorful tales about Captain Jack, and these stories appeared in Wallace's 1960 book, *Gunnison Country*. Inez Hunt and Wanetta W. Draper also interviewed people about Ellen Jack for one of their character sketches in their book *To Colorado's Restless Ghosts*. Many other sources who tell her story summarize what she wrote, adding only a few other bits of information. In the 1970s, two Wild West magazines published stories describing her exploits, one with the subtitle "She Was Born to Find Gold."

The Women's Gold Tapestry depicts Ellen Jack among the eighteen women it honors for their contributions to the settlement and development of Colorado. This nine-by-twelve-foot tapestry hangs in the Colorado State Capitol Building. Eva Mackintosh and several other women organized the project to commemorate Colorado's centennial and the country's bicentennial in 1976, and hundreds of women worked on the tapestry, with the public invited to add a few stitches to the work. The caption below the tapestry has a brief description: "Moving to Gunnison, Colorado, Mrs. Jack opened numerous successful businesses and spent much of her spare time prospecting. She was well respected by her male counterparts because of her marksmanship, business savvy, and overall tenacity."

The noted western fiction author Tony Hillerman chose a passage from Ellen Jack's autobiography for inclusion in his edited book *The Best of the West: An Anthology of Classic Writing from the American West*. Ellen's excerpt is in the section about women, followed by an excerpt from Anne Ellis. A few of the other 140 people whose writings fill the volume include Calamity Jane, Bernard DeVoto and Bret Harte. The excerpt from Ellen's autobiography described the events that occurred in Gunnison when she shot a gun out of a man's hand; she was arrested for the shooting, and a ruckus ensued in Judge David Smith's courtroom, with lawyers throwing punches at each other. Someone attributed the cleaning out of the courtroom to Ellen, which a newspaperman published with the headline "Mrs. Captain Jack, the Dare-Devil of the West." These events may have occurred in Ellen Jack's case against her assailants, but the details she wrote about are not found in other sources. As Tony Hillerman cautions in his introduction, "Authentic memory may not represent the authentic truth."

Ellen is remembered in historical exhibits and performances. Her life is described on a webpage for the National Cowboy and Western Heritage Museum. The exhibit is called *Not Just a Housewife: The Changing Roles of Women in the West*. Ellen Jack was one of 101 women featured in an exhibit from January to March 2018 at Southern Methodist University, titled *OK,*

This image is similar to the depiction of Captain Jack on the Women's Gold Tapestry. *Author's collection.*

Mrs. Captain Jack. *Author's collection.*

I'll Do It Myself: Narratives of Intrepid Women in the American Wilderness, Selections from the Caroline F. Schimmel Collection. Caroline Schimmel curated the exhibit, drawing from her collection of over twenty-three thousand narratives and representations of women in the American wilderness, including both fictional and nonfictional accounts. The Aspen Historical Society hosted a Chautauqua-style performance of Ellen's life in February 2019 in its series Aspen's Characters: Strong and Scandalous Women. The Legendary Ladies, a women's performance organization, includes Captain Jack in its *Unconventional Women of the West* show.

Captain Jack is still remembered in the Colorado Springs area. The *Cheyenne Mountain Kiva: A Journal of the Cheyenne Mountain Heritage Center* published an article in 2000 that recounted her roadhouse on the High Drive. With the popularity of Captain Jack's trail with mountain bikers, the *Springs Magazine* published an article in 2019 that informed trail users that "Captain Jack was a woman." In 2018, the Gold Camp Brewing Company named its Belgian ale Captain Ellen Jack. Given Ellen Jack's troubles over selling liquor at her roadhouse, she would probably be quite proud to have a beer named in her honor.

Captain Jack continues to be remembered as a pioneer of the Old West. As part of that tradition, stories are still told about her life as a gun-toting, colorful character, a prospector, a miner and a proprietor of her tourist roadhouse. Many of her qualities were viewed as eccentric for women during that era. She had to fight for freedoms that many people often take for granted today, such as a greater degree of economic independence for women, the ability to pursue one's own goals, the choice to marry, divorce or remain single and the ability to follow one's own inner light in religious or spiritual matters.

Captain Jack's trail passes by the location of her roadhouse on the High Drive. Lower Captain Jack's Trail starts at the Gold Camp Road, winds uphill through the Pike National Forest and arrives at the saddle that held her cabin. The large rock that Captain Jack stood in front of for many photographs is still there. If you go there, you have to use your imagination to visualize the location of her roadhouse, but if you do, you might just hear Captain Jack greet you as a tourist, offer a chicken dinner or a glass of beer and mesmerize you with tales of hidden treasure.

BIBLIOGRAPHY

Adeline J. Snevily to Ellen E. Jack, power of attorney. Book 102. Gunnison County Clerk and Recorder, March 6, 1891, 553.

Alexander Gullett and Henry L. Karr v. Ellen E. Jack. Case no. 962. Gunnison County Court, December 22, 1885.

Alexander Gullett and Henry L. Karr v. Ellen E. Jack. "Satisfaction of Judgment," book 94. Gunnison County, Colorado, 294.

Ancestry. "John T. Johnston, Missouri, Wills and Probate Records, 1766–1988, Probate Date: 3 February 1892." www.ancestry.com.

———. "Walker, Lambert R., New York, Wills and Probate Records, Vol. 28–29, 1928–1930." www.ancestry.com.

———. "Walker, Virginia, 1940 and 1950 United States Federal Census, Smithtown, Suffolk County, New York, King's Park Hospital for Insane." www.ancestry.com.

Anna Lustick v. Ellen E. Jack. Case no. 9255. El Paso County District Court, November 1912.

Aspen Daily Chronicle. "The Fire." March 28, 1889.

———. "Mining in Other Districts." August 17, 1891, 4.

Aspen Daily Times. "The Sheep Mountain Tunnel." January 31, 1895, 4.

———. "Twenty-Five Years Ago This Week." May 6, 1927.

Aspen Evening Chronicle. "Durant Street Diabolism." October 22, 1889.

Aspen Weekly Times. "John W. Adair." April 30, 1881, 2.

Babcock, N.P. "From Park Row to Early Colorado." *Scribner's Magazine*, April 1925, 377–84.

Bellevue, Beautiful View: The History of The Bellevue Valley, and Surrounding Area. Caledonia, MO: Belleview Valley Historical Society, 1983.

B.E. Shear v. Ellen Jack. Case no. 359. District Court, Pitkin County, February 16, 1886.

Bryan, Susan, and Mary Anderson. Mengoldah Lode location certificate. Gunnison County Clerk and Recorder. Book 95. December 29, 1893, 317.

Butler, Anne M. *Daughters of Joy, Sisters of Misery: Prostitutes in the American West 1865–90*. Chicago: University of Illinois Press, 1985.

Class of 1916. *Historical Sketches of Early Gunnison*. Reprinted in 1989. Gunnison, CO: Colorado State Normal School, 1916.

Colorado Daily Chieftain. "A Lynching at Montrose." October 10, 1886.

Colorado Springs Evening Gazette. "Hold Nora Gaines Whittaker Funeral Services Tomorrow." October 16, 1933.

Colorado Springs Gazette. "Carry Did No Smashing." August 25, 1906, 5.

———. "Claims She Found Cave in Mountains Near City." February 5, 1909.

———. "Completing the Bear Creek Drive." December 12, 1903.

———. "Expiration of the Great W.P.H. Lease." April 19, 1905, 6.

———. "Highwayman Worsted." October 27, 1907, 5.

———. "Murder and Suicide?" July 31, 1907.

———. "New 'Captain Jack's' Place Now Being Constructed on High Drive." April 22, 1923, 6.

———. "Threatened to Take Her Life." November 17, 1904.

———. "Woman Is Charged with Making Threats." November 26, 1904.

———. "Women Open Ore Near Town." January 29, 1904.

Colorado Springs Gazette Telegraph. "Historic Captain Jack's on Scenic High Drive Torn Down by City Crew." June 11, 1965, B1.

Colorado Transcript. "Mines Graduate Makes Good." January 11, 1906.

Crystal River Mining Company. Articles of Incorporation. Incorporation Record, Colorado, 210–13.

C.W. Young, George W. Farnham and William Elliott v. Albert A. Johnson. Case no. 2570. Colorado Supreme Court, 1889.

C.W. Young, George W. Farnham and William Elliott v. C.J.S. Hoover. Case no. 3473. Colorado Supreme Court, 1895.

Daniel J. Lehan and Elmer E. Turner to Chas H. Toll, mining deed. No. 24786. Saguache County Clerk and Recorder's Office, November 2, 1895, 469–70.

E.J. Stewart to Ellen E. Jack, mining deed. Book 78. Gunnison County Clerk and Recorder, 1885, 290.

Ellen E. Jack and H. Bryan Pearson, Juniper No. 2, location certificate. Saguache County Clerk and Recorder, October 14, 1896.

Ellen E. Jack, Juniper No. 3, location certificate. No. 26516. October 14, 1896, 567.

Ellen E. Jack, record of wills, no. J115. County Court, El Paso County, 1921.

Ellen E. Jack, Scorpio Lode, location certificate. Book 69. Saguache County Clerk and Recorder, July 26, 1894, 5.

Ellen E. Jack, Scorpion Lode, location certificate. Book 69. Saguache County Clerk and Recorder, June 13, 1894, 4.

Ellen E. Jack, Scorpion No. 2 Lode, location certificate. Book 69. Saguache County Clerk and Recorder, June 28, 1894, 4.

Ellen E. Jack, Sunday Morning Lode, location certificate. Book 69. Saguache County Clerk and Recorder, June 13, 1894, 5.

Ellen E. Jack to Adeline Jane Snevily, mining deed. Book 45. Ouray County Clerk and Recorder's Office, January 16, 1890, 215.

Ellen E. Jack v. Anna Lustick. Case no. 9661. District Court of El Paso County, Colorado, 1917.

Ellen E. Jack v. Dora W. Biebel. Case no. 2194. Gunnison County Court, August 8, 1899.

Ellen E. Jack v. John Kinkaid. Case no. 20829. District Court, Arapahoe County, April 6, 1894.

Ellen E. Jack v. Joseph Clemens, Frank Clemens, Frank Seymour and Gustav St. Germane. Case no. 163. Gunnison County Court, 1881.

Ellen E. Jack v. Josephine E. Smith. Case no. 9201. District Court, El Paso County, Colorado, 1911.

Ellen E. Jack v. Mrs. Josephine Smith. Case no. 9662. El Paso County Court, 1913.

Ellen E. Jack v. Redmond D. Walsh, decree of divorce, nullity of marriage. Case no. 1174, September 17, 1888.

Ellen E. Walsh to E.J. Stewart, mining deed. Book 78. Gunnison County Clerk and Recorder, 1885, 18.

Ellen E. Walsh v. C.F. Dunbar and J.J. Shafer. Case no. 623. Gunnison County Court, 1883.

Ellen E. Walsh v. Redmond Walsh. Case no. 391. Pitkin County Court, 1885.

Elliott, Russell R. *Servant of Power: A Political Biography of Senator William M. Stewart.* Reno: University of Nevada Press, 1983.

Find a Grave. "Johnson, Albert Augustus, Registres, Montreal, Quebec, St. James the Apostles, Anglican; Albert Augustus Johnson." www.findagrave.com.

First National Bank of Ouray v. Ellen E. Jack. Case no. 1121. Ouray County Court, December 21, 1891.

Fleming, Geo. M., and Alexander Craig to S.D. Crump, mortgage deed. Book 134. Gunnison County Clerk and Recorder, September 16, 1897, 18.

Gilbert, C.W., and Don Gilbert to W.S. Ward, mining lease, no. 70733. Gunnison County Clerk and Recorder, September 26, 1894, 523–25.

Grace, Fran. *Carry A. Nation: Retelling the Life.* Bloomington: Indiana University Press, 2001.

Gregory, Doris H. *History of Ouray: A Heritage of Mining & Everlasting Beauty.* Ouray, CO: Cascade Publications, 1995.

Guerin, E.J. *Mountain Charley.* Introduction by Fred W. Mazzulla and William Kostka. Norman: University of Oklahoma Press, 1968.

Gunnison Daily Review. November 17, 1885.

———. "Excitement on the Cebolla." January 27, 1882.

Gunnison Democrat. December 14, 1885.

———. January 16, 1886.

———. "Bill Edwards' Story." April 27, 1882.

———. "Christmas at Crystal." December 30, 1885.

———. "Gunnison's Sorrow: Death and Funeral of Judge Karr." January 29, 1889.

———. "The Sequel to the Shooting That Took Place in Cal Hayzes' Saloon Monday Night." December 22, 1881.

———. "A Sudden Death." September 22, 1888.

———. "Two Friends." January 28, 1882.

Gunnison Tribune. "Gunnison Gold Belt: O.P. Posey of the Tomboy Has Good Words for It." May 29, 1896.

———. "Gunnison Gold Belt: Thos. Tonge and Prof. Lakes Make a Careful Investigation." April 24, 1896.

———. "Mineral Hill Sold." May 24, 1895.

Hinds, Erv. *Healing the Pain of Heartache: A Physician Explores Broken Heart Syndrome*. North Charleston, SC: CreateSpace, 2010.

James Cox v. Ellen Walsh. Case no. 920. Gunnison County Court, August 1885.

John Kinkaid to Ellen E. Jack, warranty deed. Book 46. Ouray County Clerk and Recorder's Office, July 30, 1890, 212.

Josephine E. Smith v. Charles C. Smith. Case no. 7301. District Court, El Paso County, Colorado, April 3, 1903.

J.S. Purdy to Ellen E. Jack, warranty deed. Book 7. Gunnison County Clerk and Recorder, 1.

King, Joseph E. *A Mine to Make a Mine: Financing the Colorado Mining Industry, 1859–1902*. College Station: Texas A&M University Press, 1977.

Leo Erdman v. Ellen E. Jack. Case no. 1101. Ouray County Court, July 1891.

Leonard, Stephen J., and Thomas J. Noel. *Denver: Mining Camp to Metropolis*. Niwot: University Press of Colorado, 1990.

Look, Al. *Unforgettable Characters of Western Colorado*. Boulder, CO: Pruitt Press Inc., 1966.

Lucas, J. Anthony. *Big Trouble: A Murder in a Small Western Town Sets Off a Struggle for the Soul of America*. New York: Simon & Schuster, 1987.

Macon (MO) *Republican*. "From Colorado." July 23, 1885, 2.

———. "From the Far West." January 19, 1888, 3.

Malach, Roman. *White Hills: Silverado in Mojave County*. Kingman, AZ: H&H Printers, 1982.

McAdams, Dan P. *Stories We Live By: Personal Myths and the Making of the Self*. New York: William Morrow and Company Inc., 1993.

McClintock, Megan J. "Civil War Pensions and the Reconstruction of Union Families." *Journal of American History* 83, no. 2 (1996): 456–80.

National Archives. "Navy Widow's Case, Certificate no. 2177, Pensioner Ellen E. Jack, Widow of Veteran Charles E. Jack." December 2, 2017. www.fold3.com.

National Archives at Kansas City. Records of the US District Court, 21. Creator: U.S. District Court for the District of Colorado. Series Title: Bankruptcy Act of 1898 Case Files. Case: 3474 in the matter of Ellen Elliott Jack.

Neisser, Ulric, and Robyn Fivush, eds. *The Remembering Self: Construction and Accuracy in the Self-Narrative.* New York: Cambridge University Press, 1994.

New York Times. "Lambert R. Walker Dies Suddenly." December 28, 1928.

People of the State of Colorado v. Ellen E. Jack. Case no. 4569. District Court, El Paso County, Colorado.

People of the State of Colorado v. Thomas Mickey and Ellen Jack. Criminal case no. 10. Gunnison County District Court, 1881.

Polaski, Leo. *The Farm Colonies: New York City's Mentally Ill in Long Island's State Hospitals.* Kings Park, NY: Kings Park Heritage Museum, 2003.

Randall, Ruth (certified genealogist). Interview about the use of the term *aunt* when referring to Susan Bryan.

Redmond D. Walsh and Ellen E. Mickey, marriage license no. 964. Denver, Colorado, July 12, 1882.

Redmond D. Walsh and Ellen E. Walsh to John J. Donnely and S.W. Harper, quit-claim deed. Book 55. Gunnison County Clerk and Recorder's Office, 1883, 9.

Redmond D. Walsh v. Ellen E. Walsh. Case no. 531. District Court Record Book, August 31, 1885, 259–60, 268, 278–79.

R.G. Carlisle, John Tetard and Alexander Gullett, trustee v. John Gordon, as administrator of the estate of Albert A. Johnson, deceased, and Kate A. Johnson. Case no. 1805. Gunnison County Court, May 1894.

Riley, Glenda. *Building and Breaking Families in the American West.* Albuquerque: University of New Mexico Press, 1996.

Rocky Mountain News. "Attorney Crump's Home." December 11, 1904.

———. "Tapestry to Honor Role of Women in State's History." July 13, 1976.

———. "Widow Accused of Selling Liquor Will Fight." August 6, 1908.

Rocky Mountain Sun (Aspen, CO). "Mines and Mining: The Discoverer." October 28, 1882, 2.

———. "Upper Rock Creek." February 20, 1886.

Rohrbough, Malcolm J. *Days of Gold: The California Gold Rush and the American Nation.* Berkeley: University of California Press, 1997.

San Francisco Call. "Swift Murder and Suicide End Long Hunt." November 7, 1905.

San Juan Hardware Company v. Ellen E. Jack. Case no. 1112. Ouray County Court, October 5, 1891.

Sarah C. Adair v. Ellen Walsh. Case no. 458. Pitkin County Court, 1885.

Schmidt, Barbara. "We Will Confiscate His Name: The Unfortunate Case of George Escol Sellers." http://www.twainquotes.com/ColonelSellers.html.

Shores, Cyrus Wells. Letter from D.D. Fowler to C.W. Shores, December 1, 1887. Cyrus Shores Papers, Denver Public Library, Western History/Genealogy Archives.

———. *Memoirs of a Lawman*. Edited by Wilson Rockwell. Denver, CO: Sage Books, 1962.

Silbernagel, Robert. *Troubled Trails: The Meeker Affair and the Expulsion of Utes from Colorado*. Salt Lake City: University of Utah Press, 2011.

Silver, Carole G. *Strange and Secret Peoples: Fairies and Victorian Consciousness*. Oxford: Oxford University Press, 1999.

Smith, P. David. *Ouray: Chief of the Utes*. Ouray, CO: Wayfinder Press, 1987.

Solid Muldoon Weekly (Ouray, CO). "Gold Belt." December 13, 1889.

———. "Sheriff's Sale." January 15, 1892.

Sprague, Marshall. *Money Mountain: The Story of Cripple Creek Gold*. Lincoln: University of Nebraska Press, 1953.

———. *Newport of the Rockies: The Life and Good Times of Colorado Springs*. Athens: Ohio University Press, 1987.

United States, Civil Works Administration. "Tom Reilly's funeral." *History Colorado*, 1934 http://www.historycolorado.org/sites/default/files/files/Researchers/CWAPioneerInterviews_350.51-71.pdf.

United States v. Ellen E. Walsh, alias Ellen E. Jack. Case no. 480. United States District Court, District of Colorado, January 15, 1887.

Vandenbusche, Duane. *The Gunnison Country*. Gunnison, CO: B&B Printers, 1980.

Wallace, Betty. *Epitaph for an Editor: A Century of Journalism in Hinsdale and Gunnison Counties, 1875–1975*. Gunnison, CO: B&B Printers, 1987.

———. "Six Beans in the Wheel: Tales and Legends of Western Colorado." Master's thesis, Western State College of Colorado, 1956.

Wilkinson, Richard, and Kate Pickett. *The Spirit Level: Why Greater Equality Makes Societies Stronger*. New York: Bloomsbury Press, 2010.

William Elliott, George W. Farnham and C.W. Young to D.D. Fowler, lease. Book 92. Gunnison County Clerk and Recorder, June 10, 1887, 426.

William Elliott v. Alexander Gullett and Henry Karr. Case no. 768. District Court, Gunnison County, 1886.

William Elliott v. Byron E. Shear, Isaac I. Johnson and C.W. Shores. Case no. 2638. Colorado Supreme Court, 1890.

World (NY). "He Smashed Her China." May 12, 1895.

———. "Snevily Wants Revenge." May 6, 1895.

Young, Otis E. *Western Mining*. Norman: University of Oklahoma, 1970.

INDEX

ABOUT THE AUTHOR

Jane Bardal's previous publications include *Southwestern New Mexico Mining Towns* and "Oral Histories from the Grants Uranium District," in the *Mining History Journal.* She teaches psychology at Central New Mexico Community College.

Visit us at
www.historypress.com
....